W0079271

Inorganic Chemistry Concepts
Volume 10

Editors

Christian K. Jørgensen, Geneva · Michael F. Lappert, Brighton
Stephen J. Lippard, Cambridge, MA · John L. Margrave, Houston
Kurt Niedenzu, Lexington · Heinrich Nöth, Munich
Robert W. Parry, Salt Lake City · Hideo Yamatera, Nagoya

Kozo Sone, Yutaka Fukuda

Inorganic Thermochromism

With 71 Figures and 16 Tables

Springer-Verlag
Berlin · Heidelberg · New York
London · Paris · Tokyo

Prof. Dr. Kozo Sone
Prof. Dr. Yutaka Fukuda

Department of Chemistry
Faculty of Science
Ochanomizu University
Otsuka 2-1-1, Bunkyo-ku, Tokyo 112
Japan

ISBN 978-3-642-51019-9 ISBN 978-3-642-51017-5 (eBook)
DOI 10.1007/978-3-642-51017-5

Library of Congress Cataloging-in-Publication Data.

Sone, Kozo, 1923– Inorganic thermochromism.
(Inorganic chemistry concepts ; v. 10) Bibliography: p. Includes index.
1. Thermochromism. 2. Coordination compounds.
3. Chemistry, Inorganic. I. Fukuda, Yutaka, 1943–
II. Title. III. Series.
QD473.S565 1987 541.3'6 87-9563
ISBN 978-3-642-51019-9

Softcover reprint of the hardcover 1st edition 1987

2152/3020–543210

Preface

Even brilliant colors are all bound to scatter,
Who in our changing world can stay forever?
From Iroha-uta, ancient
Buddhistic poem of Japan

For many years we have been engaged in the preparation and characterization of new metal complexes and chelates, and especially the interpretation of their electronic spectra in solutions. In the course of these studies, we have encountered a number of strange changes in color which occur upon heating, cooling or compressing the solutions, or changing the nature of the solvent. Similar effects of temperature and pressure on the color were often also observed in the solid state. Records of visual observations, spectral measurements, and their interpretations and analyses accumulated each year, until we found ourselves, quite suddenly, in the middle of a fantastic world of color changes – the world of inorganic thermochromism and related chromotropic phenomena.

This book is a result of the reviews by Sone and Prof. S. Utsuno (Kagaku no Ryoiki, **22**, 222 (1968); Bunko Kenkyu, **25**, 123 (1976)), and a series of papers by Fukuda, Sone et al. published in the J. Inorg. Nucl. Chem., Bull. Chem. Soc. Japan, and various other journals after 1970. A large part of these works has been reviewed by Sone and Fukuda in the book "Ions and Molecules in Solution" of the series "Studies in Physical and Theoretical Chemistry" (Vol. 27; Ed. N. Tanaka et al., Elsevier (1983)). In addition to these works, we have naturally tried to review as many as possible related works in the literature. However, we discovered that a more or less complete review of such works is a very difficult task, since many of them are concerned with the spectacular changes in color occasionally found in the course of various types of research. They are therefore scattered over many different kinds of journals, and the phenomena of our interest are often only briefly mentioned or described, so that information cannot be gleaned from the titles or abstracts. Moreover, there exist only a few reviews of this particular field of inorganic chemistry; some of them are somewhat old, while the others are limited in scope. Thus, we realized that we were going to write the first book on this topic.

The writing of such a book has really been a challenge. Neither of us knew exactly how to accomplish such a feat; there was no "parent book" to follow, nor a "rival book" to compete with. Thus, we had to write the book in our own way. Above all, we tried to make the nature of the phenomena observed and their interpretations, that is, the beauty, interest and importance of such facts and theories in modern inorganic chemistry, easily understandable to readers of an undergraduate level, just as in our lectures, and did not fear the criticism of more advanced readers who may have wanted more sophistication. We collected, selected and arranged the works to be cited according to our own interest or taste, and tried to explain them in our own words which may sometimes be different from those of the original authors. Our aim always was to present a reasonably good coverage of important works, though a number of important papers could not be reviewed in a satisfactory way, and some even had to be left out. This is a regrettable situation, indeed, but was really inevitable in the treatment of such a wealth of material. However, we hope that the author(s) of the second book can fill in these gaps and do a better job.

In spite of all these defects, and the very fact that the color changes of many interesting systems (for example, those of $CoCl_2$ solutions; cf. Chapter B) are not yet completely understood so that their descriptions had to be qualitative, or even superficial, we have finally written this book in the present style. Now it is at your disposal; we hope, very sincerely, that you can enjoy reading it with a bit of time and patience, and find something new, interesting or useful.

We are most deeply indebted to Prof. Hideo Yamatera (Nagoya Univ.) who kindly encouraged us, in 1978, to write this book as a volume of the series "Inorganic Chemistry Concepts". His lasting support, advice and suggestions concerning the content and style of the manuscript were truly invaluable in every step of our writing. We are also cordially obliged to Dr. F. L. Boschke and Dr. R. Stumpe of the Spinger-Verlag for their friendly help and great patience, which have led us to the last page of this book after such a long time.

Our sincere thanks are also due to Prof. Masayoshi Nakahara (St. Paul's Univ., Tokyo) and Prof. Motoharu Tanaka (Nagoya Univ.), who reviewed the manuscript and supplied many valuable tips to improve the text, and to Prof. Masatoshi Fujimoto (Hokkaido Univ., Sapporo), Prof. Masayasu Mori (Osaka City Univ.), and Prof. Junzo Sunamoto (Nagasaki Univ.), for their helpful comments.

We cannot forget the hospitality of Prof. Viktor Gutmann, Prof. Ulrich Mayer, Prof. Roland Schmid and Dr. Rudolf W. Soukup (Technische Univ. Wien), shown to both of us during our visits to Vienna in 1983–1984. For their warmheartedness and friendly interest in our book we would like to say "vielen herzlichen Dank".

We would also like to thank all of our collaborators in this field, especially Dr. Tominosuke Katsurai, Prof. Michinobu Kato (Aichi Prefectural Univ. Nagoya) and Prof. Shunji Utsuno (Shizuoka Univ.) who worked with Sone on his earlier thermochromic studies, and Prof. Kenzo Nagase (Tohoku Univ., Sendai) for his recent collaboration.

Again we are thankful for the contributions made by the graduate and undergraduate students of Ochanomizu University, who worked with us in the past seventeen years. Nearly nothing could have been accomplished in our small laboratory without their support, filled with wisdom and devotion. Indeed, even this book could never have been written without them.

Tokyo, September 1987

Kozo Sone
Yutaka Fukuda

Contents

CHAPTER A

Introduction

A.I Definition of "Inorganic Thermochromism" and the Aim of this Book

The term "thermochromism" is defined and used rather loosely. Usually it implies a drastic change in color of a substance, or a system of substances, that takes place when it is heated up to a certain temperature or to a more or less well-defined temperature range. The change may be reversible, i.e., the original color may return upon cooling, or irreversible. Similar drastic changes in color, reversible or irreversible, are often observed when a substance or system is cooled down [1–3].

Some people are of the opinion that only the reversible changes should be included in the term "thermochromism". This is quite reasonable, since the irreversible changes in color caused by heating are, for the most part, merely the results of true chemical reactions, and thus do not require any special terminology. Closer studies, however, reveal that the situations are in many cases not so simple and clear-cut. For the sake of practical purposes, it seems more convenient to divide thermochromic phenomena into two groups, i.e.;

1) Reversible thermochromism, and
2) Irreversible thermochromism.

Thermochromic phenomena can be observed in solids, liquids, solutions, and gases. Numerous examples can be easily found in most textbooks of inorganic chemistry. In many cases, they are more or less closely related to other kinds of "chromisms" or "chromotropic phenomena", i.e., changes in color caused by external influences.

One of them is piezochromism, which is the change in color caused by high pressure. It has a number of similarities with thermochromism. However, reliable data are only available in limited cases, since numerous technical difficulties have been encountered in producing the high pressures (thousands and ten thousands of atmospheres) necessary to cause observable changes in color in solid or solution samples. Today, it is relatively easy to study the absorption spectra of solution samples under several thousand atmospheres of pressure. Hence we can expect much more information from this field in the near future. This phenomena may be, in fact, more common than one may expect; for example, when one grinds a colored compound in a mortar, the color sometimes changes notably in the course of grinding. This may be merely due to changes in the size and form of the particles, or to the thermochromism caused by friction. However, in certain cases it seems to be the manifestation of piezochromism itself. Careful discriminating studies of such phenomena (tribochromism) should also be of interest.

Another chromotropic phenomenon, which has been studied in a large number of cases, is solvatochromism, i.e., the change in color of a solute caused by a change in the nature of the solvent. As will be discussed later, it is most intimately related to thermochromism, since such changes in color are a direct reflection of the changes in solute-solvent (or solute-solute, or even solute-solute-solvent) interactions, which are generally influenced, more or less remarkably, by a change in temperature.

Other kinds of "chromisms", such as photochromism and electrochromism (i.e., color changes caused by light and electricity) are also known. They are interesting topics; however, at present little seems to be known about their connection with thermochromism.

It is the aim of this book to give the reader a general picture of the importance and interest in thermochromism and related chromotropic phenomena, i.e., piezochromism and solvatochromism, in the studies of inorganic chemistry, and especially in the chemistry of transition metal complexes. First of all, however, we shall discuss a few well-known examples of such phenomena in the next section, just to show their diversity and the complexity of the structural problems which very often lie behind the marvellous changes in color.

A.II Some Common Examples of Inorganic Thermochromism

A.II.1 NO_2-N_2O_4 Equilibrium in the Gas Phase

When NO_2 is sealed in a glass tube and inserted into the neck of a flask containing boiling water, the brown color of the gas evidently becomes darker. On the other hand, when the sealed tube is dipped into a beaker containing crushed ice, the color immediately fades away and remains only slightly yellowish. These changes can be repeated as often as one wishes [4].

It is now well estabilshed that these changes are due to shifts in the association-dissociation equilibrium as follows:

$$\underset{\text{Brown}}{2\,NO_2} \rightleftharpoons \underset{\text{Colorless}}{N_2O_4} \qquad \text{(Eq. A.1)}$$

NO_2 is a V-shaped molecule as shown in (a). In the valence bond (VB) picture, it is described as a resonance hybrid (b):

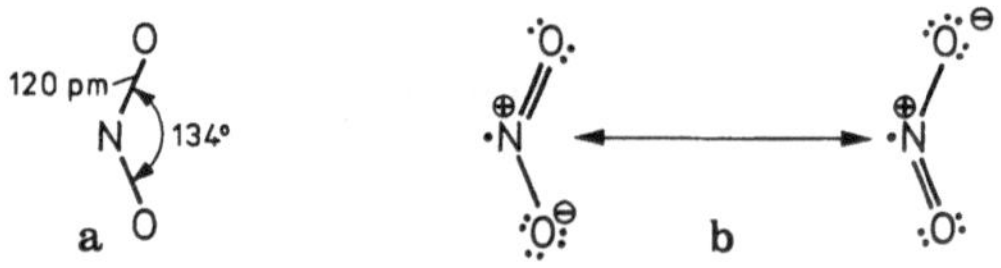

In the simplified molecular orbital (MO) picutre, one can imagine that all of the N and O atoms are sp^2-hybridized to form a σ-bond skeleton and five non-bonding sp^2 orbitals (one from N and two from each O) remain in the molecular plane. The delocalized π-bonding, non-bonding and anti-bonding MO's are shown in Fig. A.1.

The 17 valence electrons in NO_2 are distributed among these orbitals so that (i) the strongest possible bonds are formed between N and O, and (ii) as many as possible of the electrons are localized about the more electronegative O atoms. These conditions

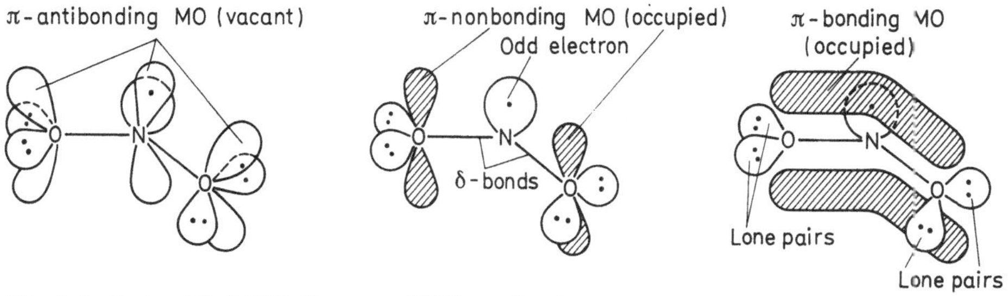

Fig. A-1. A simplified MO diagram of NO_2 molecule

are fulfilled when all of the orbitals shown in Fig. A.1 are occupied by pairs of electrons, except for the highly unstable anti-bonding orbital, and the lone pair orbital on the N atom which is singly occupied.

In both of these pictures the odd electron is located on the central N atom. The brown color of NO_2 is apparently related to the existence of this electron. It is, in fact, due to a very broad and finely split absorption band, which covers nearly the entire visible region.

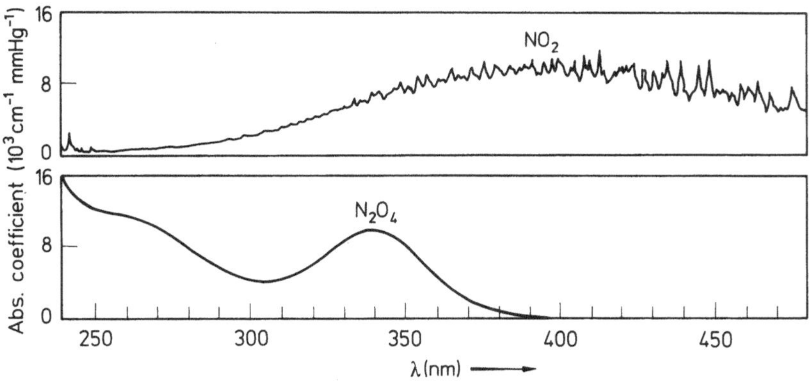

Fig. A-2. Absorption spectra of NO_2 and N_2O_4 at 25 °C. (After Hall and Blacet [5])

A part of this band is reproduced in Fig. A.2. It has a flat maximum at ca. 400 nm and extends to nearly 1150 nm. The origin of this band has been studied by many investigators, but is not yet fully understood. It is probably the result of a superposition of a number of bands, corresponding to various kinds of excitations related to the odd electron (e.g., the excitation of an electron in the non-bonding MO into the orbital of the odd electron, or the excitation of the odd electron itself to the anti-bonding orbital, etc.). On the other hand, the easy dimerization of NO_2 in the gas phase is due to this odd electron. The dimer, N_2O_4, has the shape shown below (c).

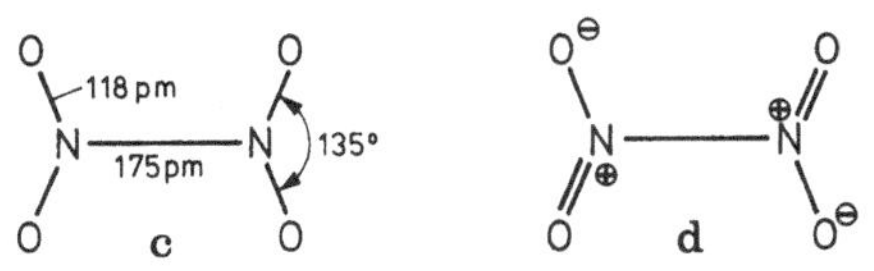

It is clear that the shape of the NO_2 moiety is nearly unchanged in the dimer; in addition, there exists a very long N—N bond, caused by the pairing of their odd electrons. Since the N atom in NO_2 is slightly but distinctly positive (dipole moment of NO_2: 0.32 Debye), the N—N bond cannot be as strong as a common single bond. This can also be seen in (d), which naturally is only one of several canocical forms. The exceptional length of the N—N bond (175 pm as compared with the standard N—N bond length of 148 pm) and the small change in enthalpy for its dissociation (57 $kJ\,mol^{-1}$ as compared with the standard N—N bond energy of 160 $kJ\,mol^{-1}$) indicate the weakness of this bond, which leads to the easy thermal dissociation of this molecule.

In spite of this very weak N—N bond, the molecule still tends to be planar, just as the isolectronic oxalate ion. Various models were proposed to explain this planarity, among which the application of the Linnet's "double quartet concept" by Bent [6] was an attractive one. Recent MO studies, however, seem to favor a model in which weak bonding interactions between the O atoms at the syn-positions are assumed [7].

It is interesting to note that even this very weak dimerization is sufficient to completely extinguish the characteristic brown color and broad band of NO_2. The dimer N_2O_4 is colorless, exhibiting only a structureless band at 340 nm and a shoulder at ca. 260 nm (similar to the spectra of NO_2^- or organic nitro compounds). This drastic spectral change, connected with the very easy interconversion of N_2O_4 and NO_2, brings about the observed thermochromism.

It may be added here that this thermochromism does not only take place in the gas phase, but also in the liquid [8, 9] and in solutions [10]. In the solid state, all of the NO_2 molecules are in the dimeric form. However, they already begin to dissociate at the melting point (−11.2 °C), becoming a pale yellow liquid containing ca. 0.01% NO_2. This value increases by nearly a factor of ten in going to the boiling point (21.15 °C), where the liquid is reddish brown. The dissociation continues to increase in the gas phase, until it reaches a maximum at 140 °C. The same equilibrium is also observed in certain organic solvents. However, it was found that the equilibrium constants of the dimerization given in (Eq. A.1) determined in these solvents are much larger than those found in the gas phase, depending on the polarity of the solvent (ca. 10^3 times larger in cyclohexane and CCl_4, and somewhat less in CH_3CN). The gas phase equilibrium is naturally shifted to the N_2O_4 side by an increase in pressure, so that one can say that the NO_2-N_2O_4 mixture is not only a thermochromic system, but also a solvatochromic and a piezochromic one.

Cl_2 and Br_2 are also thermochromic; their colors fade away upon cooling, becoming colorless and pale yellow at −195 °C [12a]. Br_2 and I_2 vapors are piezochromic. Their spectral changes in the ultraviolet region indicate that they form weak dimers, $(X_2)_2$, upon compression, which are held together by van der Waals' forces [10a–10c]. Chromotropic phenomena in gases are thus by no means rare.

A.II.2 Color of Iodine in Various Solutions

It is well known that the color of iodine in solution strongly depends on the nature of the solvent, i.e., iodine is highly solvatochromic in solutions [11]. In a non-polar solvent, such as an aliphatic hydrocarbon, CCl_4 or CS_2, the color of the solution is violet. Its visible absorption spectrum exhibits a band at ca. 500 nm, the shape and intensity of which are not very different from that of iodine vapor. However, in more polar solvents such as alcohols, acetone, ether or pyridine, or in π-electron-rich solvents such as aromatic hydrocarbons or their derivatives, a strong band appears in the ultraviolet region and the original band at ca. 5000 nm is shifted towards shorter wavelengths. The color of the solution thus becomes yellow to brown in the former case, and red of various shades in the latter. The curves in Fig. A.3, which were taken from Benesi and Hildebrand [12], show some examples of these spectral changes.

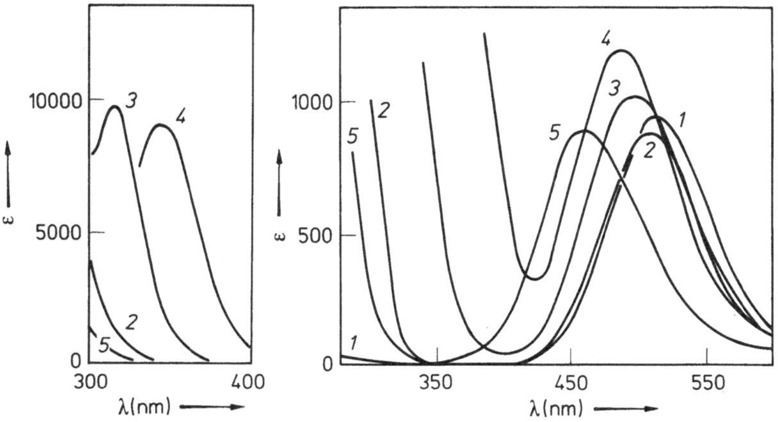

Fig. A-3. Absorption spectra of I_2 in various solvents. Curves *1* to *5* correspond to solutions of I_2 in CCl_4, $C_6H_5CF_3$, C_6H_6, $C_6H_3(CH_3)_3$ and $(C_2H_5)_2O$, respectively. (After Benesi and Hildebrand [12])

A closer examination of such curves and the spectral changes caused by the stepwise addition of these solvents to the violet I_2 solutions in non-polar solvents reveal that a special kind of solvation of the I_2 molecules, called "charge transfer (CT) complex formation", takes place in these non-violet colored solutions. For example, in the yellow to brown solutions, this solvation involves the formation of a coordinate bond from a donor atom of the solvent (D:) to I_2, leading to a resonance between the two structures (a) and (b) shown below:

$$\underset{a}{\text{D:}---\text{I—I}} \longleftrightarrow \underset{b}{\text{D:}\rightarrow\text{I}---\text{I}}$$

On the other hand, the solvation in the red solutions involves a corresponding resonance between (a′) and (b′), in which π-electrons are partially transferred from a solvent molecule to I_2.

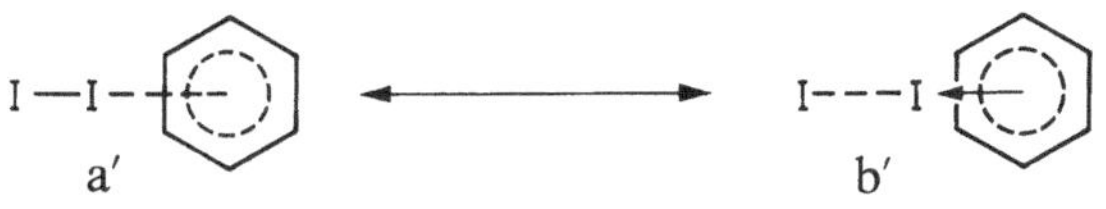

a′ b′

In this case one can assume (a) or (a′), in which the component molecules are just in contact with one another, represents the main structure of the complex, and the coordinate-bonded structure (b) or (b′) is resonating with it to a lesser extent. With such resonances, one can expect an excited state of the complex, in which (b) or (b′) is the main structure and (a) or (a′) is resonating with it to a lesser extent. A characteristic, usually intense, absorption band will then appear in its electronic spectrum, which corresponds to a net transfer of charge from the solvent to I_2.

The strong absorption of the non-violet solutions of I_2, which appears in the ultraviolet region, can be assigned to this "CT" band. This can be seen, for example, from a comparison of the curves 2, 3 and 4 in Fig. A.3. It is clear that the observed CT bands are shifted towards the red (lower energy) in the order of $C_6H_5CF_3 \rightarrow C_6H_6 \rightarrow C_6H_3(CH_3)_3$, i.e., in the order of increasing electron density on the benzene ring, or in the order of increasing π-electron donor ability of the solvent.

It can also be seen that the band near 500 nm is shifted towards the blue (higher energy) in the same order. This is easy to understand, since this band is related to the excitation of a "lone pair" (more exactly speaking, π^*) electron on I_2 into a vacant σ^* orbital. In the course of complex formation, this same orbital is used to accept the electron pair from the solvent. Hence, an excitation to it becomes more difficult, as the CT complex-forming interaction increases [12a].

Similar comparisons may be made with the spectra in ether (curve 5) and other polar solvents, in which the CT complexes are mainly the result of σ-type interactions. Here, however, the situation is often more complicated; if the solvent is a strong σ-donor (i.e., Lewis base), the I—I bond in the complex is weakened, and easily split in the following way:

$$D: --\rightarrow I—I \longrightarrow D: \rightarrow I^+ + I^-$$

For example, this occurs in pyridine, in which a pair of donor molecules are coordinated to I^+, forming a complex $[I(py)_2]^+$. In a resent study, Aronson et al. [13] stated that about 18% of the iodine in a 1.5 M solution of I_2 in pyridine is in the ionized state. Partial ionization also takes place in other solvents. The I^- that is formed may combine with I_2 molecules to form $I_3{}^-$ or higher polyhalide ions.

Apart from such complications, it is now interesting to note that certain organic solvents, such as ethyl stearate or oleate, form distinctly thermochromic solutions of I_2. At room temperature they are brown, but when heated (above 80 °C in the above two solvents) they become violet. The reason for this thermochromism is now apparent; at room temperature the polar ends of the long chain molecules form CT complexes with I_2, which easily dissociate upon heating. The liberated I_2 molecules are then surrounded by the non-polar chains of the solvent, just as in aliphatic hydrocarbon solutions. The same kind of thermochromic change is also observed in a non-polar solvent containing a small amount of a polar or aromatic solvent. Spectrophotometric studies on such phenomena have often been performed to determine the stability of the CT complex of I_2 with the latter solvent [11].

The blue solution obtained in the familiar "iodine-starch reaction" is also typically thermochromic. When the solution is warmed a little, the blue color fades away quickly and the solution becomes almost colorless. Upon cooling, the blue color returns. This change has been known for a very long time, probably since 1812 when this blue color first attracted the interest of chemist [15]. Strangely enough, even today this color change is by no means fully understood.

The structure of the blue I_2-starch complex, which was first suggested by Freudenberg [14] and supported by a number of physical studies carried out by Rundle et al. [15], is truly fantastic [16]. Starch is composed of two kinds of glucose polymers, amylose and amylopectine. The former is a long chain polymer, while the latter is a branched one. Only the former can form the typical blue complex with I_2. Exactly speaking, its long chain is by no means straight, but is randomly coiled in solution. It can sometimes form a helix with a long channel or "chimney" passing through it. Now the inner diameter of such a chimney (500 pm) is only slightly larger than the diameter of an I_2 molecule measured perpendicular to the I—I axis, so that the chimney can accommodate, by means of intermolecular attraction, a large number of I_2 molecules in it, as shown in Fig. A.4.

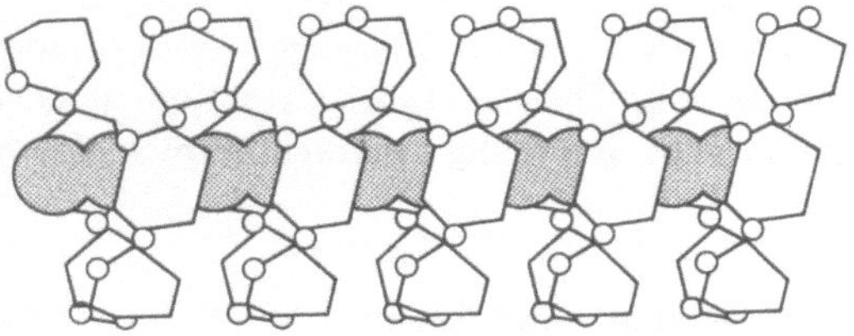

Fig. A-4. Model of I_2-starch complex. Note that, according to recent studies [16], the amylose chain is screwed into a left-handed helix as shown here, with 6 glucose units per pitch (800 pm). The mean I–I distance in the polyiodide chain is 310 pm. (After Rundle et al. [15] with correction)

These iodine molecules are thought to be in close contact with one another, so that they form a very long chain-like polymer themselves, in which the outermost electrons of the I_2 molecules are to some extent delocalized over the entire chain, similar to the π-electrons of polyenic compounds. The high mobility, or excitability, of these electrons leads to the strong blue color of the complex. Figure A.5 shows a typical spectrum for such a blue solution. A very strong absorption band at ca. 600 nm, covering the entire visible and near-ultraviolet regions, appears upon formation of the complex

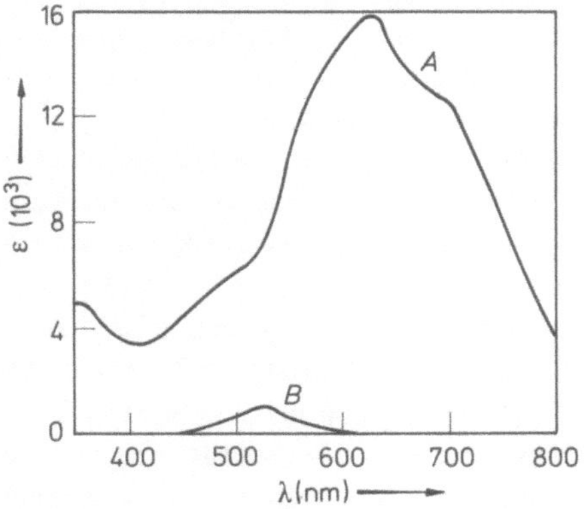

Fig. A-5. Absorption spectra of I_2: A, 5×10^{-5} M in 0.01% potato starch solution; B, 1×10^{-3} M in CCl_4. (After Rundle et al. [15])

Although this picture seems to be quite reasonable, the true situation is far more complex with many structural enigmas still unsolved. For example, recent Mössbauer and Raman studies by Teitelbaum et al. [18], and X-ray studies by Noltemeyer and Saenger [17], have shown that the particles in the chimney are not merely I_2 molecules, but mainly I_5^- (I—I---I$^-$---I—I or I—I—I$^-$---I—I), or a complex (or complexes) of iodine with I^- ions. The role of the I^- ion in this complex is not clear.

Possibly it creates a certain electron distribution along the chain, which stabilizes the chain in the chimney and increases the mobility of electrons in the chain. Our understanding of the true nature of the starch-iodine complex, and its blue color, is thus quite imperfect at present.

The thermochromism of the blue solution is, however, very spectacular, if we imagine it on the atomic scale. When the solution is heated slightly, all of the I_2 molecules and the I^- ions leave the chimney and diffuse into the aqueous phase. A large part of the helix, which owes its stability to the interaction with the I_2 molecules, then collapses and changes into a random coil. Since the color of I_2 in water is light brown, a dilute solution is nearly colorless. When the solution is cooled, the scattered molecules and ions surround the random coil. Their mutual interaction quickly restores the structure of the helix and the chimney. It is really a marvellous play of ions and molecules.

A.II.3 Ruby and Some Compounds of Mercury and Silver

Everyone knows that a ruby is a red gem, but in fact it is a thermochromic gem, changing its color with temperature [19].

Ruby is a solid solution of Cr_2O_3 in α-Al_2O_3. These two oxides are of the same structure with a corundum-type lattice. However, the lattice constants of the former are somewhat larger (a = 496 pm, c = 1360 pm for Cr_2O_3 vs. a = 476 pm, c = 1299 pm for α-Al_2O_3), corresponding to the larger size of Cr^{3+}. Common red rubies contain less than 8 (atomic)% of Cr. In these crystals the lattice constants of α-Al_2O_3 remain nearly unchanged. When the Cr content is increased above 8%, the lattice constants are also increased linearly. The crystals become violet and finally green, which is the color of pure Cr_2O_3. Rubies are reversibly thermochromic, changing first to violet and then to green upon heating (Fig. A.6). The change occurs at around −183 °C, 187 °C and 377 °C when the Cr content is 58%, 8% and 2%, respectively [18]. This is clearly the result of thermal expansion, which makes the mixed crystal more "Cr_2O_3-like" when hot. Similar thermochromic "artificial rubies" can be prepared by dissolving Cr_2O_3 in crystals of other trivalent oxides, such as those of Ga, La-Ga and Y-Al.

These changes in color can be reasonably understood in terms of the ligand field theory of transition metal complexes. It is well known that the colors of the transition metal compounds are due to their so-called d—d bands, or electronic transitions between the d-orbitals split under the electric field of the ligands. A Cr^{3+} ion in a ruby is, so to speak, being squeezed into a octahedral cage of O^{2-} ions which is too small. The resulting ligand field is too strong for a $[Cr^{III}O_6]$-type complex, which is usually violet (as in chromium alum with $[Cr(H_2O)_6]^{3+}$), blue or green (as in $[Cr(urea)_6]^{3+}$). The unusually large splitting of the d-orbitals caused by such a field is the origin of the red color of common ruby. When it is heated and expanded, the Cr^{3+} ions become more relaxed, regaining their proper color of violet to green, at the expense of the Al^{3+} ions which are now in ligand cages which are too big. The competition between these two cations to be in a proper-sized cage is thus the origin of this thermochromism. Spectral data obtained by McClure [20] are given in Fig. A.6.

It is interesting to note that ruby is also piezochromic. According to Minomura and Drickamer, the absorption band at 18150 cm^{-1} (551 nm), which corresponds to the magnitude of the d-orbital splitting (10 Dq), shifts towards lower wavelengths upon

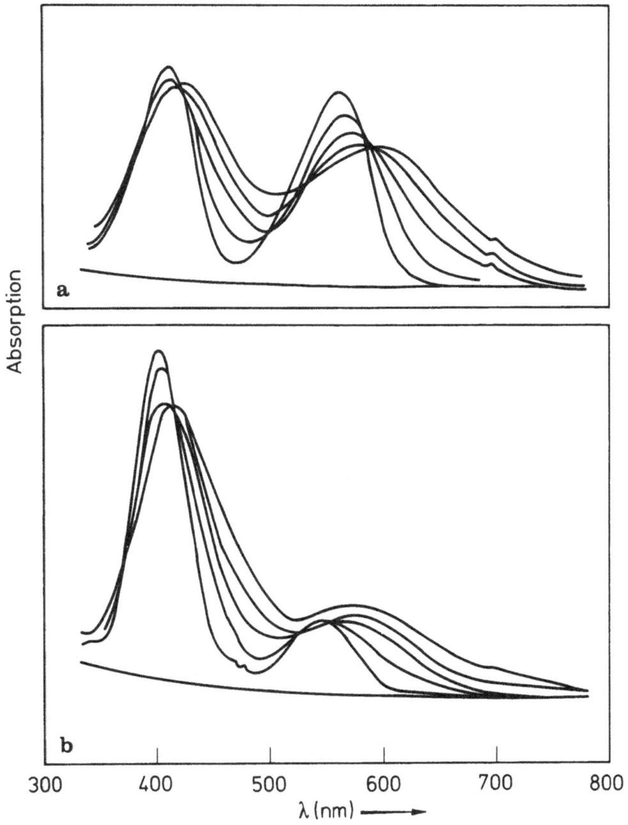

Fig. A-6 a, b. Absorption spectra of ruby (Cr^{3+} content: ca. 0.2 mole%) at various temperatures. **a:** with light polarized ⊥ c; **b:** with light polarized // c. From left to right, the five curves correspond to the spectra at ca. 25, 220, 440, 650, and 900 °C, respectively. (After McClure [20])

compression of the crystal. It reaches ca. 19150 cm^{-1} (522 nm) at 100 kbar [21]. This is what we would have expected from the above discussion [22].

Thermochromic changes, which are essentially similar to those of ruby, can also be observed when common colored salts of transition metals are cooled in liquid nitrogen. The green hydrated salts of Ni^{2+}, containing $[Ni(H_2O)_6]^{2+}$ ions, become sky-blue, while the red salts of $[Co(H_2O)_6]^{2+}$ change to a yellow color [23]. These changes evidently correspond to an increase in the ligand field, due to a strong contraction of the crystals.

Another example of solid state thermochromism among common inorganic materials is that of red tetragonal HgI_2 caused by a phase transition. It becomes reversibly yellow at 127 °C, changing into a rhombic high-temperature modification. These colors are due to a strong absorption band in the ultraviolet region, which is probably a ligand to metal (L → M type) CT band, i.e., a band arising from the partial transfer of an electron from the I^- ion to Hg^{2+}, the tail of which extends into the visible region. The deformation of this band accompanying the phase transition causes the apparent change in color.

In this case, the following resonance comes into effect.

$$\underset{a}{I^{-}---Hg^{2+}} \longleftrightarrow \underset{b}{I^{0}---Hg^{+}}$$

A highly deformable anion such as I^-, in contact with a metal ion of variable oxidation number (a), can easily lose one of its electrons to the metal ion (b). The result is such a resonance and a related CT absorption as in the case of donor-iodine complexes.

Similar changes are also observed in some double salts of HgI_2, even at much lower temperatures. For example:

$$Ag_2HgI_4\text{: Yellow} \underset{}{\overset{50.7\,°C}{\rightleftharpoons}} \text{Orange} \qquad Cu_2HgI_4\text{: Red} \overset{67.0\,°C}{\rightleftharpoons} \text{Black}$$

The change in color of the latter compound is very drastic and may involve the enhancement of a charge transfer of $M \rightarrow L \rightarrow M'$ type, i.e., a strong shift of a CT band corresponding to a flow of electrons from Cu^+ to Hg^{2+} towards longer wavelengths [2]:

$$Cu^{+}----I^{-}---Hg^{2+} \xrightarrow{h\nu} Cu^{2+}---I^{-}----Hg^{+}.$$

AgI is also remarkably thermochromic [24]. At room temperature it consists of pale yellow crystals of wurtzite structure, with tetrahedrally coordinated Ag^+ and I^- ions. When the compound is heated, the yellow color becomes more and more intense until finally a dark red solid exhibiting remarkable electric conductivity is obtained. This is believed to be due to the fact that above ca. 147 °C the Ag^+ ions begin to escape from their original cages of four I^- ions, and to randomly migrate into other vacant cages (in a wurtzite structure, only one half of the tetrahedral cages (or holes) formed by the I^- ions are occupied by the Ag^+ ions). The CT band of $L \rightarrow M$ type, which again is in the ultraviolet region, is strongly deformed and is shifted into the visible region by the disorder, bringing about the observed color change.

It may be added that many oxo-salts of Ag(I) are colored. Some of them are also thermochromic. Table A.1 shows how their colors, which are again ascribed to a CT band of $L \rightarrow M$ type, depend on the strength of Ag-O interaction [24]. Among them, Ag_3PO_4 is known to become orange-red upon heating. This change is related to the anisotropic thermal vibration of Ag^+, by which the length of the Ag-O bond tends to diminish. Many other thermochromic changes are known; they seem to exist among the compounds of large and soft ions.

Table A.1. Color and Ag-O bond length in the oxo-salts and oxide of Ag(I) (After Wells [24])

Compound	Ag-O length (pm)	Color
$AgClO_3$	251	Colorless
Ag_2SO_4	250	Colorless
$KAgCO_3$	242	Colorless
Ag_3PO_4	234	Yellow
Ag_3AsO_4	234	Deep red
Ag_2CO_3	230	Yellow
Ag_2O	206	Brown-black

The three examples given in this section were taken, more or less at random, from the vast number of thermochromic systems encountered in inorganic chemistry. However, the reader will now recognize the fact that they appear in many variations, among metallic and nonmetallic compounds in any state, and in connection with many kinds of structural changes which are often hard to understand completely.

By far the largest number of these phenomena are known among the heavy metal complexes and chelates; especially in their aqueous and non-aqueous solutions. A typical and very well-known example, i.e., the thermochromic, piezochromic and solvatochromic behavior of cobalt(II) chloride slutions, will now be treated in more detail in the next chapter.

In general, thermochromic changes are those with $\Delta H^\circ \neq 0$, and piezochromic changes are those with $\Delta V \neq 0$. The observed changes in color become more apparent with the increase of ΔH° or ΔV°, if changes of the same type are considered. These simple criteria are often useful in studying these phenomena.

References

1. Day, J.H.: Chem. Rev. **63**, 65 (1963)
2. Day, J.H.: Chem. Rev. **68**, 649 (1968)
3. Day, J.H.: "Kirk-Othmer's Encyclopedia of Chemical Technology", Vol. 6, Wiley-Interscience, New York (1979), p. 129
4. For a review on this topic, cf., e.g., Bell, C.F.: "Syntheses and Physical Studies of Inorganic Compounds", Pergamon Press, Oxford (1972), p. 35
5. Hall, Jr., T.C., Blacet, F.B.: J. Chem. Phys. **20**, 1745 (1952)
6. Bent, H.A.: Inorg. Chem. **2**, 747 (1963)
7. Alrichs, R., Keil, F.: J. Am. Chem. Soc. **96**, 7615 (1974); Howell, J.M., Van Wazer, J.R.: ibid. **96**, 7902 (1974)
8. James, D.W., Marshall, R.C.: J. Phys. Chem. **72**, 2693 (1968)
9. Vosper, A.J.: J. Chem. Soc. **A1970**, 2192
10. Redmond, T.T., Wayland, B.B.: J. Phys. Chem. **72**, 1626 (1968)

10a. Passchier, A.A. et al.: J. Phys. Chem. **71**, 937 (1967)

10b. Passchier, A.A., Georgy N.W.: J. Phys. Chem. **72**, 2697 (1968)

10c. Tamres, M. et al.: J. Phys. Chem. **72**, 966 (1968)

11. For a review on this topic, and on similar "CT complexes" in general, cf. Briegleb, G.: "Electronen-Donator-Acceptor-Komplexe", Springer-Verlag, Berlin (1961); Andrews, L.J., Keefer, R.M.: "Molecular Complexes in Organic Chemistry", Holden-Day, San Francisco (1964); Foster, R.: "Organic Charge-Transfer Complexes", Academic Press, London (1969); Yarwood, J.: "Spectroscopy and Structure of Molecular Complexes", Plenum Press, London (1973)
12. Benesi, H.A., Hildebrand, J.H: J. Am. Chem. Soc. **71**, 2703 (1949)

12a. Cf. e.g. Greenwood, N.N., Earnshaw. A.: "Chemistry of the Elements", Pergamon Press, Oxford (1984)

13. Aronson, S. et al.: J. Phys. Chem. **86**, 1035 (1982)
14. Freudenberg, K. et al.: Naturwiss. **27**, 850 (1939)
15. Rundle, R.E. et al.: J. Am. Chem. Soc., **66**, 2116 (1944)
16. Saenger, W.: Naturwiss. **71**, 31 (1984)
17. Noltemeyer, M., Saenger, W.: J. Am. Chem. Soc. **102**, 2710 (1980)
18. Teitelbaum, R.C. et al.: J. Am. Chem. Soc. *100*, 3215 (1978)
19. Thilo, E. et al.: Naturwiss. **37**, 399 (1950)
20. McClure, D.S.: J. Chem. Phys. **36**, 2757 (1962)
21. Minomura, S., Drickamer, H.G.: J. Chem. Phys. **35**, 903 (1961)
22. Some data on the thermochromic and piezochromic spectral changes of minerals are collected in Burns, R.G. (Ed.): "Mineralogical Applications of Crystal Field Theory", Chap. 9, Cambridge Univ. Press (1977)

23. Jørgensen, C. K.: "Absorption Spectra and Chemical Bonding in Complexes", Pergamon, Oxford (1962), p. 145
24. Wells, A. F.: "Structural Inorganic Chemistry", 3rd Ed., p. 267, Oxford Univ. Press (1967)

CHAPTER B

Thermochromic and Chromotropic Phenomena of Cobalt(II) Chloride Solutions and Related Systems

B.I Introductory Remarks: Visual Observations on Aqueous and Alcoholic Solutions of Cobalt(II) Chloride

The pink color of an aqueous solution of cobalt(II) chloride is known to be due to that of the 6-coordinate octahedral hexaaqua cation, $[Co(H_2O)_6]^{2+}$. However, anhydrous $CoCl_2$ is blue and its solutions in various organic solvents, such as alcohols or acetone, are also blue. In general, the blue color is ascribed to the formation of tetrahedral 4-coordinate complexes. These are often formed when the Co-L (L: ligand) bond is rather weak and ionic, and the ligand is bulky (hence, interligand repulsion favors a tetrahedral geometry). For example, the spectral and conductometric studies of a number of investigators (e.g. Katzin, Libuś, Osugi, and their collaborators [14, 17, 31]) have indicated that the blue neutral species of the composition $[Co(Solv)_2Cl_2]$ predominates in ethanol and higher alcohols (Solv = a solvent molecule; cf. Chap. B.II).

$$\underset{\text{Blue}}{CoCl_2} \xrightarrow{H_2O} \underset{\text{Pink}}{[Co(H_2O)_6]^{2+}} + 2Cl^- \qquad \text{(Eq. B.1)}$$

$$\xrightarrow{ROH} \underset{\text{Blue}}{[Co(Solv)_2Cl_2]}$$

In most cases one says that "a 6-coordinate Co(II) complex is pink and a tetrahedral 4-coordinate one is blue", but there are a number of exceptions. First of all, anhydrous $CoCl_2$ possesses a $CdCl_2$ structure, so that the Co(II) ion in it is 6-coordinate. It is, however, deep blue (still more strangely, when cooled with liquid air the color of this compound changes into a dull pink without a structural change [1]). On the other hand, the chelate $[Co(dipm)_2]$ (dipm$^-$ = dipivaloylmethanate anion, $(CH_3)_3C{-}CO{=}CH{-}CO{-}C(CH_3)_3^-$) is tetrahedral 4-coordinate, but is pink [2]. There are moreover many complexes of intermediate colors. A careful spectral comparison and other structural data (if available) are needed in such cases in order to determine the geometry of the complexes in question.

The pink color of an aqueous solution becomes gradually and reversibly reddish violet upon heating. This change can be especially clearly observed, when the concentration is relatively high (e.g. 0.1 – 1 M; 1 M = 1 mol dm^{-3}) and the temperature is near 100 °C. On the other hand, the blue color of an ethanolic solution or that of the solutions in higher alcohols does not change significantly upon heating. When we add a small amount of water, drop by drop, into such a blue solution, it first becomes bluish violet, then reddish violet, and finally pink. We can thus easily guess that the pink and blue species coexist in the violet solutions and that the equilibrium between them is shifted "from blue to pink" by increasing the amount of water in solution. Expressed in a simplified form, the equilibrium will be:

$$\underset{\text{Blue}}{[Co(EtOH)_2Cl_2]} + 6H_2O \rightleftharpoons \underset{\text{Pink}}{[Co(H_2O)_6]^{2+}} + 2Cl^- + 2EtOH. \quad \text{(Eq. B.2)}$$

Various intermediate species containing two or even three kinds of ligands (EtOH, Cl^- and H_2O) in various proportions will coexist with these extreme species, so that the true equilibrium is much more complex.

It is now interesting that the violet solutions of cobalt(II) chloride in EtOH-H_2O mixed solvent are very much thermochromic; upon heating they become blue and upon cooling in ice they become pink. These solutions, which are sometimes called "liquid thermometers" and can be used as an attractive classroom demonstration for chemical equilibrium, indicate that the formation of the blue 4-coordinate complexes (e.g. $[Co(EtOH)_2Cl_2]$) in such a mixture is endothermic, while the formation of the pink 6-coordinate species (e.g. $[Co(H_2O)_6]^{2+}$) is exothermic. By changing the solvent in the original blue solution (e.g. from ethanol to propanol etc.) and by controlling the amount of water added, it is possible to construct "thermometers" which change their colors about certain temperatures.

Since methanol is intermediate in structure between water and ethanol, one can expect a methanol solution of $CoCl_2$ to behave like an ethanol-water solution. In fact, this is the case; the former solution is reddish violet and strongly thermochromic just like the latter, even without the addition of any water.

A note on the works of Katsurai may be of some interest here. As referred to above, even an aqueous solution of $CoCl_2$ is thermochromic if the concentration and temperature are high enough. However, the change in color which takes place is not as drastic, and as a result is rather difficult to observe. To overcome this difficulty, Katsurai developed a unique technique which he called "visual autoclaving" [3–6], and carried out a number of observations on the colors of aqueous solutions of heavy metal salts and complexes heated above 100° C.

In his experiments, various colored solutions were sealed into heat-resistant glass tubes and placed obliquely on a circular holder, which, similar to a small merry-go-round, was slowly driven by a motor over an electric heater. One of the tubes was open and contained silicone oil in which a thermometer was inserted. All of the tubes were uniformly heated up to (or sometimes above) ca. 200 °C. The changes occuring in each tube were observed visually or photographed in color. The set was placed in a large box with safety glass windows and a fine gauze ceiling covered with a large thick towel to protect the observer from possible explosion. Some of the results he obtained in this way are listed below:

(1) The change in color of a $CoCl_2$ solution caused by heating becomes much more apparent above 100 °C. The solution becomes notably violet at 200 °C. The solutions of other Co(II) salts, such as the bromide, sulfate and nitrate, are much less thermochromic, but all show similar color changes when heated to 200 °C. Solutions of the corresponding Ni(II) salts are all nearly the same green below 100 °C, but at 200 °C they become yellowish. In both cases, the ease of the color change increases in the order:

Nitrate < Sulfate ≈ Bromide < Chloride.

Solutions of $CuCl_2$ are known to change from blue to green by heating up to 100 °C. This color change is also strongly enhanced above 100 °C whereby the solution becomes yellowish green.

All of these changes in color are reversible and are thought to be due to the formation of halogeno- sulfato-, and nitrato-complexes of these metals at high temperatures. On the other hand, $CuSO_4$ is strongly hydrolyzed above 100 °C to form a white precipate (probably a basic sulfate).

(2) In contrast to the solutions of simple salts described in (1), the colors of the solutions of the edta chelates of these metals remain unchanged under the same conditions. The thermal stability of the edta chelate is also demonstrated by the fact that the Fe(III) chelate remains unchanged at 200 °C, while the simple salts of Fe(III), and even the cyano-complexes of Fe(II) and Fe(III), are easily hydrolyzed at the same temperature to form hydroxides or oxides. However, Hg(II) chelate was found to decompose upon heating and deposit fine drops of mercury.

(3) Characteristic thermochromic changes were also observed in solutions of $H[AuCl_4]$, $K[AuCl_4]$, $H_2[PtCl_6]$, $K_2[PtCl_6]$ and UO_2Cl_2 heated above 100 °C.

The merit of Katsurai's method is its simplicity. It is highly valuable as a preliminary test for a more quantitative approach. We can propose a number of hypotheses concerning the state of aqueous solutions above 100 °C. For example, we can imagine that water above 100 °C will lose much of its polarity to become something like a polar organic solvent, since the interaction of its molecules among themselves and with the charged solute particles will be notably diminished by their enhanced thermal motion. Its dielectric constant (D) will therefore be much lower than that at room temperature. In fact, it is 55.7 at 100° C, according to the equation relating D and temperature t [°C]

$$D = 87.740 - 0.40008t + 9.398 \times 10^{-4}t^2 - 1.410 \times 10^{-6}t^3,$$

proposed by Kortüm [7]. Ionic association in water above 100°C will thus be favored, and the water of hydration about the ions will easily be driven out by thermal agitation. Therefore, we can expect that reactions such as

$$[M(H_2O)_6]^{2+} + nX^- \rightleftharpoons [M(H_2O)_yX_n]^{(2-n)+} + (6-y)H_2O \quad \text{(Eq. B.3)}$$

will proceed to the right hand side, causing thermochromic changes as described in (1). Such high-temperature aqueous solutions are an entirely unknown field; it seems that even now there are only a few works aimed at a more quantitative approach to such systems, although Franck once studied the spectra of cobalt(II) and nickel(II) salt solutions at very high temperatures and pressures [8, 9] and Giggenbach constructed a very compact high-temperature-high-pressure spectrophotometric cell and studied reactions involving the formation of the blue supersulfide ion S_2^- [10–13]. The application of such sophisticated techniques to other systems, such as Katsurai's, will certainly bring about many fruitful results.

B.II Low-Temperature Spectrophotometric Studies on Alcoholic Solutions of Cobalt(II) Chloride

The absorption spectra of aqueous and organic solutions of $CoCl_2$ have been studied by many investigators, such as Katzin and Gebert [14], Cotton and Goodgame [15], Libuś [16–19], Buffagni and Dunn [20], Fine [21], and Gutmann and his collaborators [22–25]. Most of the studies were carried out at room temperature. Often the concentration of the solute and added chlorides were changed. The relations between the spectra and the structures of the complex species formed in solition, and, in particular, the stepwise formation of the complexes with increasing number of Cl^- ions, have been elucidated more or less quantitatively in many cases.

In these studies, use was made of the fact that there is a clear-cut difference between the spectra of the octahedral cobalt(II) complexes and those of the tetrahedral ones. Figure B.1 illustrates two typical examples, e.g., the visible absorption band of the pink aqueous solution of $CoCl_2$ due to $[Co(H_2O)_6]^{2+}$ ion, and that of the deep blue complex $[CoCl_4]^{2-}$ which is formed when there is a large excess of Cl^- ions in the solution.

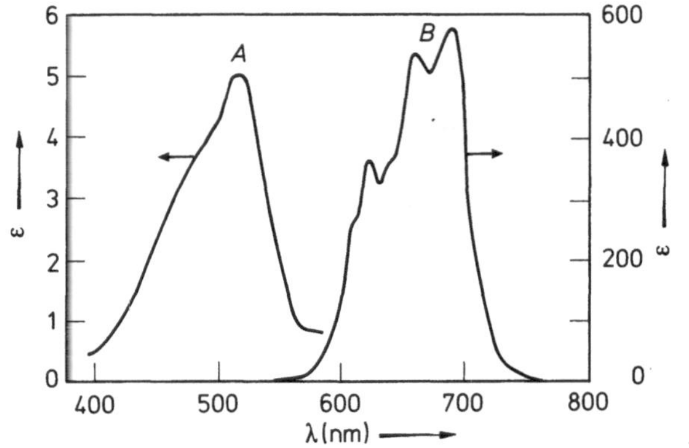

Fig. B-1 a, b. Typical absorption spectra of **(a)** octahedral $[Co(H_2O)_6]^{2+}$ and **(b)** tetrahedral $[CoCl_4]^{2-}$ in the visible region. (Partly after Cotton and Wilkinson [26])

Although exceptions do exist, excitation of the electronic system in an octahedral complex, which is usually a high spin one, leads to a weak band ($\varepsilon \lesssim 10$) near 500 nm. In a tetrahedral complex a much stronger band ($\varepsilon \gtrsim 100$) at 600–700 nm, which is split in a complicated way is obtained. An additional band appears in the near-infrared spectra of these complexes, the position and appearance of which also depends on the stereochemistry of the complex. These spectral features have served very well in the structural studies of many cobalt(II) complexes.

In the last section, we pointed out that the blue color of an ethanolic solution of $CoCl_2$ does not change significantly upon heating. The same can be said about the solutions in higher alcohols, while methanolic solutions are very highly thermochromic. There is now one more question: "What will happen, when we cool these solutions?"

Upon cooling these solutions in a dry ice-ether mixture their blue color fades away and, in certain cases, becomes quite pink, when the temperature is as low as −50 °C. A quantitative analysis of these changes was carried out by Nieuwpoort et al. [27] using a unique cell with circulating cooling fluid. In four C_1–C_3 alcohols they confirmed that the intensity of the characteristic "tetrahedral band" at ca. 660 nm notably decreases upon cooling and that the shape of the spectral curve approaches that of an octahedral complex.

Sone et al. [28] also carried out the same kind of measurements with a simpler home-made cryostat, using eight C_1–C_4 alcohols, among which only tert-BuOH could not be used due to its high freezing point (25 °C). Their data are similar to those of Nieuwpoort et al., but they pointed out several features of this thermochromism which have not been reported before. Some of their data on the methanolic, ethanolic and sec-BuOH solutions are given in Fig. B.2.

The data on the ethanolic solutions given in Fig. B.2b are the simplest to understand. They indicate disappearance of the tetrahedral species $[Co(EtOH)_2Cl_2]$, which predominates at room temperature, and the formation of an octahedral species

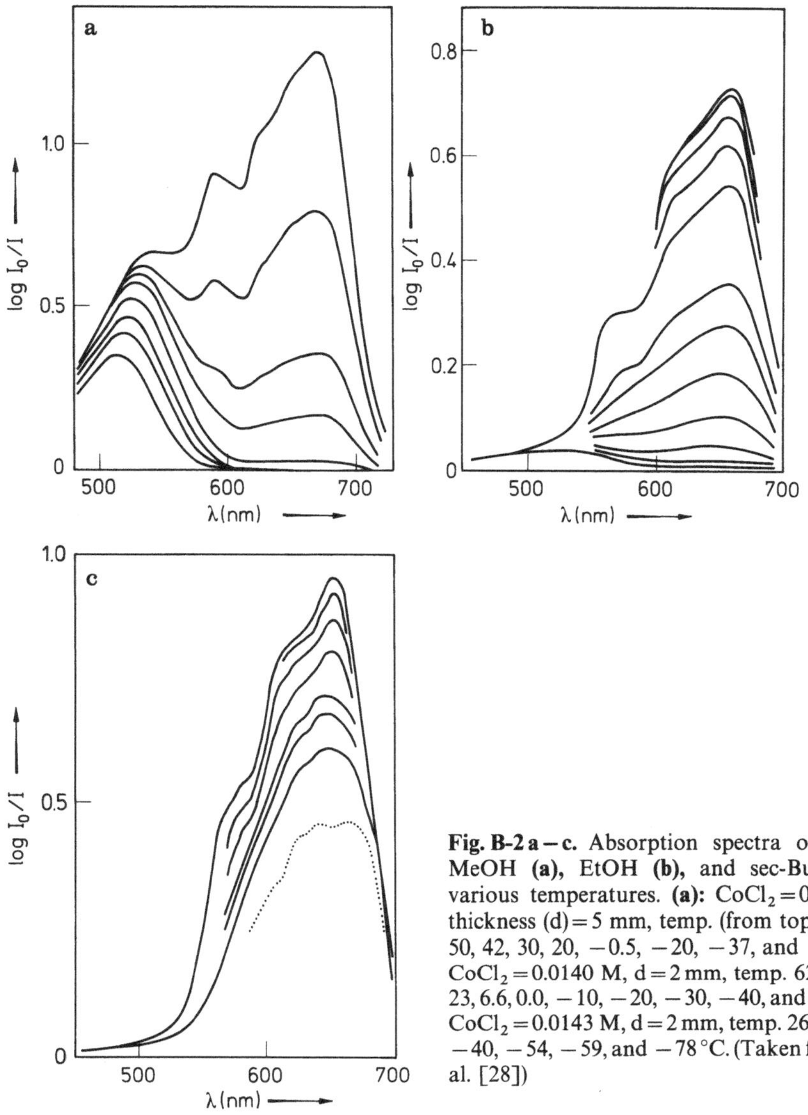

Fig. B-2a–c. Absorption spectra of $CoCl_2$ in MeOH **(a)**, EtOH **(b)**, and sec-BuOH **(c)** at various temperatures. **(a)**: $CoCl_2 = 0.103$ M, cell thickness (d) = 5 mm, temp. (from top to bottom) 50, 42, 30, 20, −0.5, −20, −37, and −70 °C; **(b)**: $CoCl_2 = 0.0140$ M, d = 2 mm, temp. 62, 54, 45, 34, 23, 6.6, 0.0, −10, −20, −30, −40, and −55 °C; **(c)**: $CoCl_2 = 0.0143$ M, d = 2 mm, temp. 26, −20, −30, −40, −54, −59, and −78 °C. (Taken from Sone et al. [28])

instead. The spectrum of the latter is quite similar to that of $[Co(H_2O)_6]^{2+}$ shown in Fig. B.1, suggesting that the species that is formed is $[Co(EtOH)_6]^{2+}$. However, the spectrum may also be due to $[Co(EtOH)_5Cl]^+$, since $[Co(H_2O)_5Cl]^+$, formed to some extent in aqueous solutions of high Cl^- concentrations, is known to exhibit a spectrum similar to that of $[Co(H_2O)_6]^{2+}$ (with somewhat higher λ_{max} and ε_{max} values: 529 nm and 13 compared with 510 nm and 5 for the latter [29]; the spectrum of the ethanolic solution at −55 °C is closer to that of $[Co(H_2O)_5Cl]^+$). The thermochromic equilibrium in ehtanolic solution can therefore be expressed as follows:

$$[Co(EtOH)_2Cl_2] + (3\text{–}4)EtOH \rightleftharpoons ([Co(EtOH)_5Cl]^+ \text{ or } [Co(EtOH)_6]^{2+}) + (1\text{–}2)Cl^-, \quad \text{(Eq. B.4)}$$

This equation represents the gradual dissociation of the predominant neutral species $[Co(EtOH)_2Cl_2]$ at lower temperatures, which finally results in its complete conversion into an octahedral ionic species.

The situation is apparently somewhat different in methanol. Figure B.2a again shows the disappearance of the blue, tetrahedral species upon cooling. However, the disappearance occurs at much higher temperature, and the λ_{max} and ε_{max} of the remaining octahedral species still continue to decrease upon further cooling. This fact indicates that at least two octahedral species exist in equilibrium. They can be taken to be $[Co(MeOH)_5Cl]^+$ and $[Co(MeOH)_6]^{2+}$, respectively; the latter is favored at lower temperatures, since their spectra are quite similar to those of $[Co(H_2O)_5Cl]^+$ and $[Co(H_2O)_6]^{2+}$, respectively.

It is interesting to note that the conductivity and spectral data of Osugi and Kitamura (cf. Chap. B.III) [31, 33] also indicate the existence of these two species in dilute methanolic solutions of $CoCl_2$ at room temperature.

On the other hand, the value of λ_{max} for the band of the blue species (670 nm) is somewhat higher than those observed in ethanol or higher alcohols (655–660 nm). Its shape is also distinctly different. In their pioneering study in 1950, Katzin and Gebert [14] already indicated that the tetrahedral species causing this type of band is not $[Co(MeOH)_2Cl_2]$, but $[Co(MeOH)Cl_3]^-$. The data on the spectra of $[Co(Solv)_2Cl_2]$ and $[Co(Solv)Cl_3]^-$-type complexes in other solvents (acetone, DMSO, etc.) also seem to support their view [21, 22, 24]. The thermochromic equilibrium in methanol can therefore be formulated in terms of two successive steps, i.e.:

$$[Co(MeOH)Cl_3]^- + 4MeOH \underset{heat}{\overset{cool}{\rightleftharpoons}} [Co(MeOH)_5Cl]^+ + 2Cl^- \qquad \text{(Eq. B.5a)}$$

and

$$[Co(MeOH)_5Cl]^+ + MeOH \underset{heat}{\overset{cool}{\rightleftharpoons}} [Co(MeOH)_6]^{2+} + Cl^-. \qquad \text{(Eq. B.5b)}$$

It may be added that $[Co(MeOH)Cl_3]^-$ is not the predominant species in methanol, in contrast to higher alcohols in which $[Co(Solv)_2Cl_2]$ generally predominates at room temperature. At room temperature the band at 670 nm is much weaker than the one at 530 nm (Fig. B.2a). Since the band of a blue, tetrahedral Co(II) complex is usually more than ten times stronger than that of a pink, octahedral one (cf. Fig. B.1), we can conclude that the blue complex $[Co(MeOH)Cl_3]^-$ exists in a negligible amount in this solution, while most of the dissolved $CoCl_2$ ($\gg 90\%$) is ionized into $[Co(MeOH)_5Cl]^+$ and Cl^-. The thermochromic change in methanol is therefore due to the complete dissociation of this small amount of tetrahedral species and the further ionization of the predominant octahedral species, both of which are favored by an even slight amount of cooling.

The situation shown in Fig. B.2 for a sec-BuOH solution (c) is peculiar. The spectrum observed at room temperature is very similar to that of the ethanolic solution, showing that the predominant blue species is $[Co(sec\text{-}BuOH)_2Cl_2]$. The band of this species is weakened upon cooling, but at much lower temperature than in ethanol. Again, it seems that the original tetrahedral species is gradually converted at lower temperature into an octahedral species with a much weaker absorption band.

However, such a band is difficult to recognize even at the lowest temperatures studied, since it is still under the tail of the strong tetrahedral band.

On the other hand, the shape of the band of the remaining blue species changes remarkably at lower temperatures, becoming very broad and structureless. A comparison with the spectral data in other solvents [17–25, 35] indicates that, along with the tetrahedral-octahedral changes, a part of the $[Co(sec\text{-}BuOH)_2Cl_2]$ is being converted into another tetrahedral species with an additional halide anion, $[Co(sec\text{-}BuOH)Cl_3]^-$.

The mode of the spectral changes accompanying this conversion can be inferred from the study of Libuś et al. [17], in which a large number of spectra of $CoCl_2$ in iso-PrOH, with various amounts of Cl^- added, are given and analyzed. Since the thermochromic behaviour of $CoCl_2$ in this solvent is very similar to that in sec-BuOH (see below), this approach seems to be quite convincing. The data of Sawada et al. [35] for such a system in acetone are also useful for a comparison, since the spectral changes involved are quite similar. From these and similar data one can readily predict that, when a fraction of the $[Co(sec\text{-}BuOH)_2Cl_2]$ is converted into $[Co(sec\text{-}BuOH)Cl_3]^-$, the overlap of the curves of both species will make the main peak much broader and smear the fine structure of the shorter-wavelength side, as observed in the low-temperature spectra of Fig. B.2c.

The features of the thermochromism in sec-BuOH can thus be expressed as follows: the equilibrium of the type given in Eq. B.4 is shifted to the right hand side upon cooling, but with much difficulty, and a large part of the released Cl^- recombines with the original blue complex, forming a new "blue to blue" equilibrium:

$$\underset{\text{Blue}}{[Co(sec\text{-}BuOH)_2Cl_2]} + Cl^- \rightleftharpoons \underset{\text{Blue}}{[Co(sec\text{-}BuOH)Cl_3]^-} + sec\text{-}BuOH \qquad \text{(Eq. B.6)}$$

which hinders the appearance of the pink color. Although cooling in sec-BuOH favors the formation of ionic species, the Co^{2+} ions thus tend to avoid an increase in their coordination number as far as possible.

It is interesting to note that the spectral changes observed in other primary alcohols, such as n-PrOH, n-BuOH and iso-BuOH, are all quite similar to those in

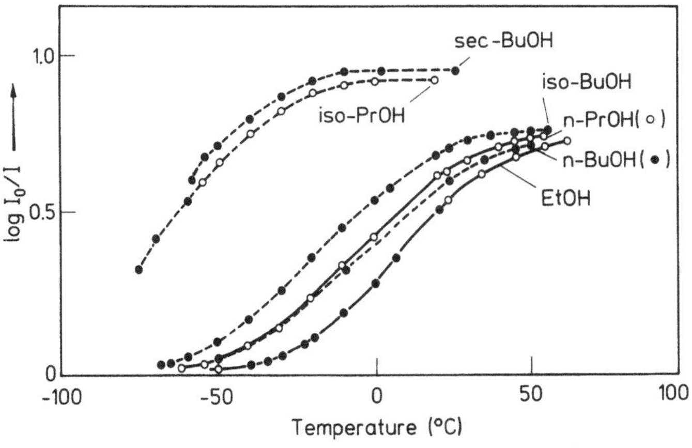

Fig. B-3. The relation between the intensity of the band of the blue species (measured at 660 nm) and the temperature for various alcohols. Concentration of $CoCl_2$: 0.0140 M, cell thickness = 2 mm. (Taken from Sone et al. [28])

ethanol, while those in another secondary alchohol, iso-PrOH, are like those in sec-BuOH. These similarities can be most clearly seen in Fig. B.3. The relationships between the apparent height of the band of thetrahedral species and the temperature are shown in this figure. The curves for the primary alcohols, and the two curves for the secondary alcohols, lie close together.

The ease of the (blue → pink) or (tetrahedral → octahedral) change for $CoCl_2$ thus lies in the following order:

$$\underbrace{\text{MeOH} \gg \text{EtOH} > \text{n-PrOH} \approx \text{n-BuOH} > \text{iso-BuOH}}_{\text{Primary alcohols}} \gg \underbrace{\text{iso-PrOH} > \text{sec-BuOH}}_{\text{Secondary alcohols}} \qquad \text{(Eq. B.7)}$$

The drastic qualitative and quantitative changes in going from methanol to other primary alcohols, and from the latter to secondary alcohols, are clear. These changes are much larger than those found in going, for example, from ethanol to n-PrOH or from iso-PrOH to sec-BuOH, i.e., those caused by the elongation of the carbon chain. The thermochromic behavior of $CoCl_2$ in an alcohol is thus determined by the number of carbon chains attached to the OH-bearing carbon atom.

More insights into the meaning of this interesting situation can be found in comparing these data with those on various other chromotropic phenomena of $CoCl_2$ solutions, which will be described in the following sections.

B.III Effects of Water, High Pressure and Dilution on the Spectra of Alcoholic Solutions of Cobalt(II) Chloride

It is interesting to note that the order given in (Eq. B.7) for the ease of the (blue → pink) or (tetrahedral → octahedral) change of $CoCl_2$ is valid for a number of other chromotropic phenomena. They indicate that tertiary alcohols should appear at the end of this order, so that the entire order should be as follows:

$$\begin{aligned} &\text{MeOH} \gg \text{other primary alcohols} \gg \text{secondary alcohols} \\ &> (\text{or} \gg) \text{ tertiary alcohols} \end{aligned} \qquad \text{(Eq. B.8)}$$

The existence of this order was first noted by Kato et al. [36], who studied the effect of the addition of a small amount of water to these solutions. They found that, when water is added little by little to the blue alcoholic solutions, the color of the solutions in methanol and primary alcohols fades away quickly, leaving a pink solution. More water is needed to extinguish the blue color of solutions in secondary alcohols, and still much more in tertiary alcohols. Their spectral data, some of which are shown in Fig. B.4, evidently confirm the existence of the order given in (Eq. B.8) in this "water effect".

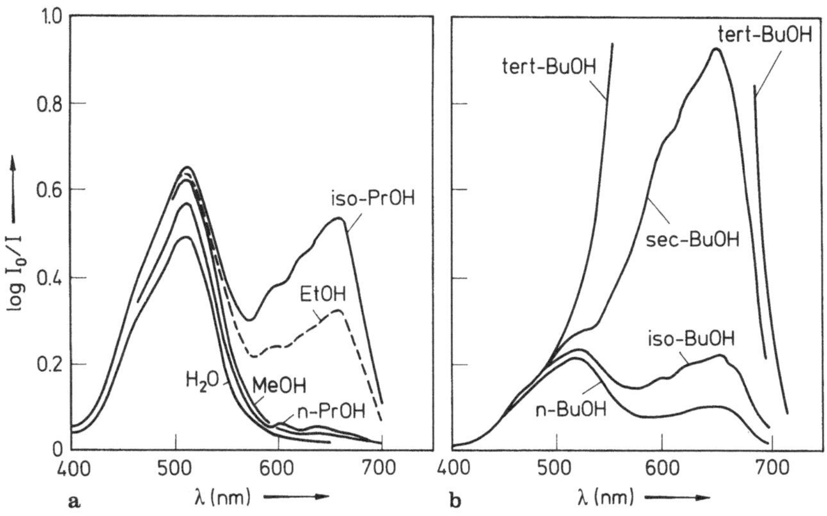

Fig. B-4 a, b. Effects of added water on the spectra of alcoholic solutions of $CoCl_2$, **(a)**: $CoCl_2$ $=0.1$ M, water 10% (vol.), temp. 14 °C; **(b)**: $CoCl_2=0.03$ M, water 4% (vol.), temp. 26 °C. In **(a)**, one cannot compare the curves of n- and iso-PrOH directly with those of MeOH and EtOH, since the PrOH solutions contain more water when the concentration is expressed in molar fractions. Correction for this makes the curve of n-PrOH approach that of EtOH, while the band of the blue species in iso-PrOH is much stronger. (Taken from Kato et al. [36])

If we express this color change in a simple way, i.e., using the equations

$$[Co(MeOH)Cl_3]^- + 6H_2O \rightleftharpoons [Co(H_2O)_6]^{2+} + 3Cl^- + MeOH(\text{in MeOH}) \qquad \text{(Eq. B.9a)}$$

and

$$[Co(Solv)_2Cl_2] + 6H_2O \rightleftharpoons [Co(H_2O)_6]^{2+} + 2Cl^- + 2Solv(\text{in other alcohols}) \qquad \text{(Eq. B.9b)}$$

and neglect any possible intermediates, the results of Kato et al. may be explained as follows. Among the blue species formed in alcoholic solutions of $CoCl_2$, those in methanol and primary alcohols are most easily attacked by H_2O and dissociated into their components. Those in other alcohols are attacked with more difficulty, which increases from secondary to tertiary alcohols.

Another interesting observation was made by Osugi and Kitamura [30–33], who studied the effects of high pressure (between 1×10^3 and 1×10^4 kg cm^{-2}) on the absorption spectra of alcoholic solutions of $CoCl_2$. Some of their results are given in Fig. B.5.

As can be seen, the blue color of the solutions in methanol, ethanol and other primary alcohols at room temperature fades away at higher pressures. The mode of the spectral change is very similar to that observed upon cooling. In this case the band of the blue species in methanol nearly disappears at 2×10^3 kg cm^{-2}, while in ethanol (and other primary alcohols) this band is still stronger than that of the pink species at the same pressure and only disappears at a higher pressure (ca. 5×10^3 kg cm^{-2} or

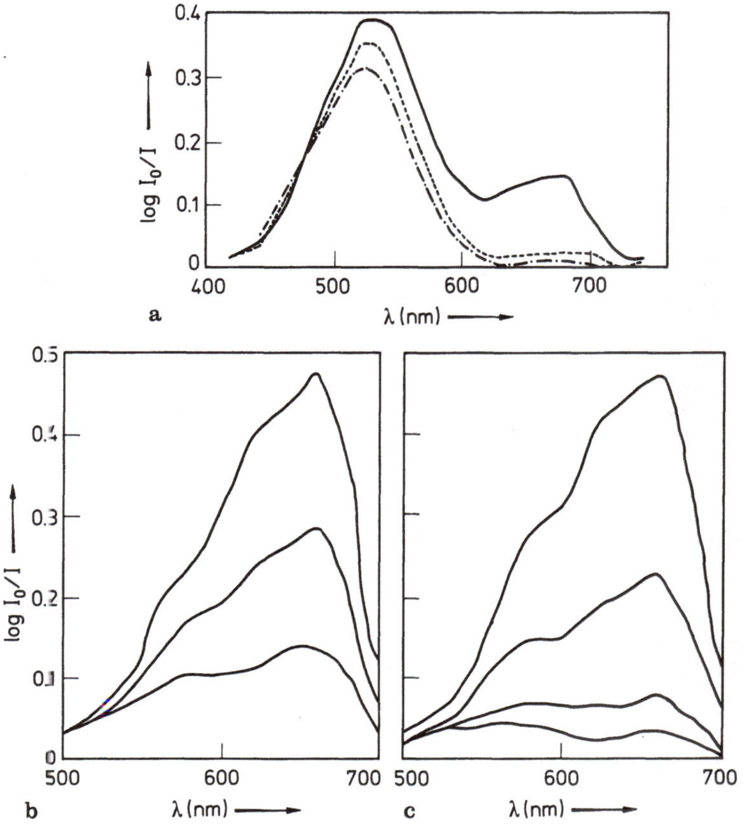

Fig. B-5 a – c. The effects of pressure (P) on the spectra of $CoCl_2$ in MeOH **(a)**, EtOH **(b)** and n-BuOH **(c)**. Conditions: **(a)**: $CoCl_2 = 0.1$ M, P (from top to bottom, kg cm^{-2}) = 1, 2×10^3 and 4×10^3, **(b)**: $CoCl_2 = 8.8 \times 10^{-3}$ M, P = 1, 1×10^3, and 2×10^3; **(c)**: $CoCl_2 = 6.1 \times 10^{-3}$ M, P = 1, 2×10^3, 4×10^3, and 5×10^3. Cell thickness: 2.5–3.0 mm. All of curves have been corrected for the concentration increase caused by compression. (Taken from Osugi and Kitamura [30])

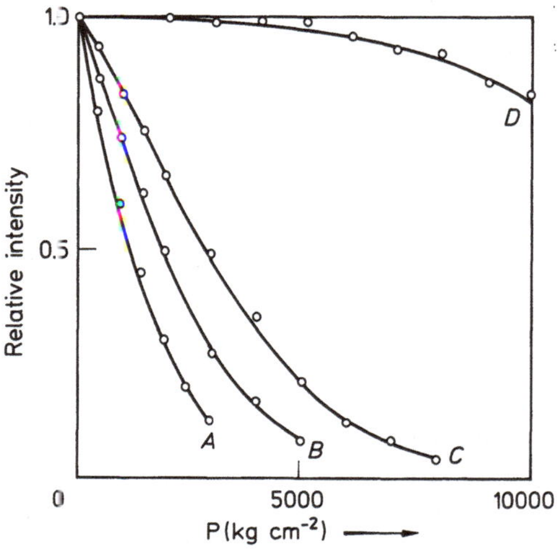

Fig. B-6. Relationship between the band intensity at 660 nm and the pressure (P) in various solvents: A = EtOH, B = n-BuOH, C = iso-BuOH, D = sec-BuOH. (Taken from Osugi and Kitamura [30])

above). On the other hand, the blue color in secondary alcohols, such as iso-PrOH and sec-BuOH, fades away at still higher pressures, so that even at 1×10^4 kg cm^{-2} a large part of the blue species remains in solution. Figure B.6 shows these relations for four alcohols, among which A—C are primary alcohols, while D is a secondary alcohol.

Solutions in tert-BuOH could not be studied because the solvent solidified at 240 kg cm^{-2}. The spectrum shows practically no change at this pressure, indicating that a change in color is very hard to produce, probably more so than that in secondary alcohols. All of these data confirm the order given in (Eq. B.8) for this phenomenon, i.e., piezochromism.

Osugi and Kitamura also found that the difficulty of producing a change in color of the blue species formed in secondary and tertiary alcohols is also observed by merely diluting the solutions containing them [28, 29]. Figure B.7 shows some of the spectra observed. The solutions in primary alcohols deviate strongly from Beer's law, while those in secondary (and tertiary) alcohols obey it. The spectral changes observed in primary alcohols are again very similar to those observed by cooling or compressing the same solutions.

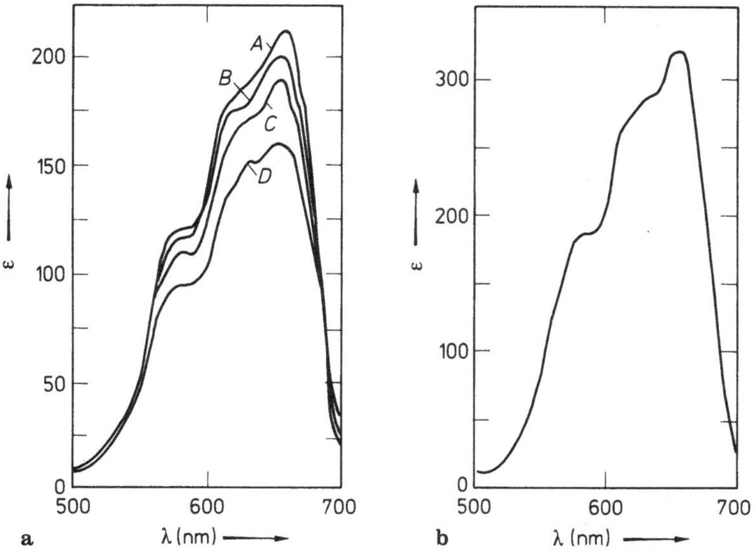

Fig. B-7 a, b. Effects of dilution on the spectra of $CoCl_2$ in n-PrOH **(a)** and iso-PrOH **(b)**. Concentrations (in 10^{-4} M, from *A* to *D*) in **(a)** are 7, 5, 3, and 1, respectively; the spectrum in **(b)** is independent of concentration from 5×10^{-4} M to 0.5×10^{-4} M. Temp. = ca. 25 °C. (After Kitamura [33])

In analyzing their data, Osugi and Kitamura came to the conclusion that the large differences between primary and secondary alcohols are due to the fact that the octahedral species, $[Co(Solv)_6]^{2+}$ or $[Co(Solv)_5Cl]^+$, are much more unstable in secondary alcohols than in primary alcohols, which in turn is due to the mutual steric hindrance of the bulky solvent molecules packed in the coordination sphere.

The idea of Osugi and Kitamura may be expressed as follows. For example, with butanols one can easily imagine that the long alkyl chains in $[Co(n\text{-}BuOH)_6]^{2+}$ are able to move around the

Co^{2+} ion like the legs of an octopus, so that they still retain much of their freedom of motion (i.e., entropy). On the other hand, the bulky alkyl groups in $[Co(sec\text{-}BuOH)_6]^{2+}$ must adequately accomadate each other, somewhat like the pieces of a zigsaw puzzle, to share the limited space around the small Co^{2+} ion. Much of their entropy is lost, thus destabilizing the entire system. Even when the alkyl groups are packed in this manner, their mutual repulsions will hinder the close approach of the ligand atoms to Co^{2+}, so that the system will also be energetically destabilized. The equilibrium of the type shown in (Eq. B.4) is thus strongly shifted to the left hand side with sec-BuOH, compared to n-BuOH. A similar difference will exist between n-PrOH and iso-PrOH. Tert-BuOH shifts the equilibrium further to the left, while methanol, which is the most free from such troubles, favors the (blue → pink) change more so than any other alcohol.

The effect of dilution can readily be understood from this point of view; as to the piezochromism, it can be explained as follows. In general, the application of a high pressure shifts an ionization equilibrium, such as the one given in (Eqs. B.4 or B.5) to the right hand side, since the ionic species that are produced attract solvent molecules more strongly than the original neutral (or less charged) species and reduce the total volume of the solution (Le Chatelier's principle). However, with secondary or tertiary alcohols, both the instability of the octahedral ionic species and the difficulty of compactly arranging the bulky solvent molecules in their outer spheres oppose the effects of the outer pressure, so that the observed change in color is very much smaller. The result is the order given in Eq. B.8.

We can now apply the same point of view to the problems encountered in low-temperature thermochromism (Chap. B.II). When these solutions are cooled, the solvent molecules lose their thermal energy. Their apparent polarity increases and favors the ionization of the solute (cf. Chap. B.I). Moreover, the volume of the solution decreases and favors the formation of ionic species. All of these effects are again opposed by the difficulty of formation of the octahedral ionic species in higher alcohols, so that the same order (Eq. B.8) is valid for the color change.

Applying the ideas of Osugi and Kitamura, much of the chromotropic behavior of $CoCl_2$ in alcohols can thus be explained in at least a qualitative way. There are, however, other factors which must be taken into acount. One of them, which probably is of importance, is the solvation of the Cl^- ions formed together with the octahedral cations. It is plausible that they are strongly solvated with methanol or primary alcohols, but more weakly solvated with secondary or tertiary alcohols, changing just in the order of (Eq. B.8).

This trend can be seen most simply by comparing the acceptor numbers (AN; cf. Chapter D) of these alcohols, which are a measure of their ability to combine with anions dissolved in them. Some of them are as follows:

Alcohols:	MeOH	EtOH	n-PrOH	n-BuOH	iso-PrOH	tert-BuOH
AN:	41.5	37.9	37.3	36.8	33.6	27.1

It is of interest, in this connection, to note that their donor numbers (DN; cf. Chapter C), which are a measure of their ability to combine with cations (e.g. Co^{2+}), are not so different (19.1 for MeOH and 19.6 for n-PrOH).

This effect must also be a driving force for the order (Eq. B.8), observed in so many chromotropic phenomena. A strong solvation of Cl^- shifts all of the equilibria, e.g. (Eq. B.4), to the right hand side, while a poor solvation causes a shift in the opposite direction. Moreover, in the latter case, the Cl^- ions that are formed are more unstable and reactive; this explains why they can attack the $[Co(ROH)_2Cl_2]$ complex to convert it into $[Co(ROH)Cl_3]^-$, as shown in the thermochromism of sec-BuOH solutions (Eq. B.6).

Other factors, involving various polarity characteristics of the alcohols and the modes of their changes with temperature or pressure, are also expected to influence the phenomena. In the case of the "water effect", the situation is further complicated; the relative coordination tendencies of the alcohols, Cl^- and H_2O, and the relative stabilities of the tetrahedral and octahedral species in various alcohol-water mixtures, have to be compared, so that the number of factors to be considered is indeed large.

It is thus plausible that the order given in (Eq. B.8) is brought about by a number of such conceivable factors. However, more quantitative approaches to this problem are generally quite difficult. Although many investigators tried to obtain thermodynamic parameters for the equilibria involved, it seems that a large portion of their values are apparent ones obtained with various simplifications, and are not useful enough to distinguish the effects of different factors. Sometimes the results are inconsistent with each other; at other times they cannot be compared with each other with sufficient reliability.

Apart from the technical difficulties involved and a lack of accurate fundamental data on some of the alcohols, an important reason for such a situation is the fact that all of the equilibria involve many intermediate species. Their neglect makes the results unreliable. To take all of the possibilities into account will necessitate a large number of measurements of high accuracy and their perfect analysis, which would really be a formidable task even with the use of computerized instruments. These are probably the reasons why these color changes still remain as one of the unsolved problems of inorganic chemistry.

B.IV Thermochromic Equilibria of Cobalt(II) Chloride in Aqueous Solutions

B.IV.1 General Remarks and "Inert Salt Effect"

It has already been mentioned that aqueous solutions of $CoCl_2$ are also thermochromic, when they are concentrated and hot enough. When an excess of Cl^- ions is added to such a solution at room temperature, at first a violet and then a blue color appears owing to the formation of various chloro complexes. The equilibria involved in these changes have been studied by many investigators. Here the results of the spectrophotometric studies by Bjerrum et al. will be considered [37a].

Figure B.8 shows the distribution curves of the original hexaaqua cation and various chloro complexes formed with an increase in Cl^- concentration. It can easily be seen that the hexacoordinated monochloro complex, $[Co(H_2O)_5Cl]^+$, is readily formed in a relatively dilute HCl solution. Of the higher chloro complexes, the di- and trichloro complexes are formed only in limited amounts. Therefore, above ca. 10 M of HCl, the tetrachloro complex $[CoCl_4]^{2-}$ becomes the predominant species.

The fact that $[Co(H_2O)_5Cl]^+$ and $[CoCl_4]^{2-}$ are the main species formed in aqueous solutions of high Cl^- concentrations can also be seen from data of Zeltman et al. [37] and Libuś [37b], who measured the NMR and visible spectra of $CoCl_2$, either with the addition of large amounts of Cl^- ions, or in very concentrated solutions. Libuś obtained similar results for $CoBr_2$ solutions. However, she could only confirm the formation of $[Ni(H_2O)_5Cl]^+$ and $[Ni(H_2O)_5Br]^+$

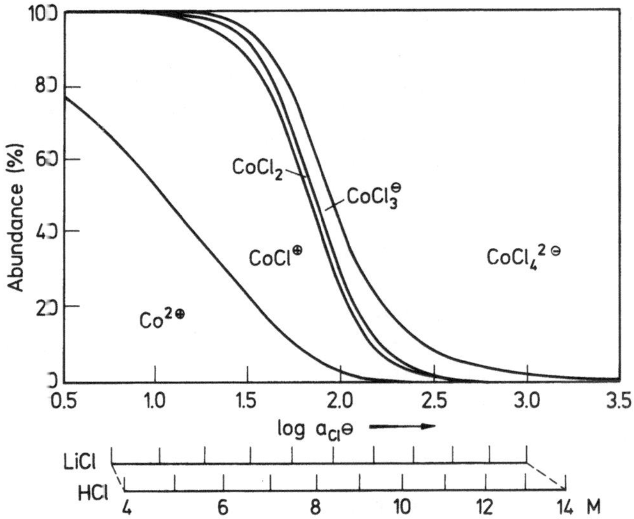

Fig. B-8. Formation of $CoCl_n^{(2-n)+}$ complexes in aqueous solutions of increasing Cl^- activity (a_{Cl^-}). Coresponding concentrations of LiCl and HCl are also given. Among the complexes shown, Co^{2+}(=$[Co(H_2O)_6]^{2+}$), $CoCl^+$ (and probably $CoCl_2$) are red to red-violet and octahedral, while $CoCl_3^-$ and $CoCl_4^{2-}$ are blue and tetrahedral. (After Bjerrum et al. [37a])

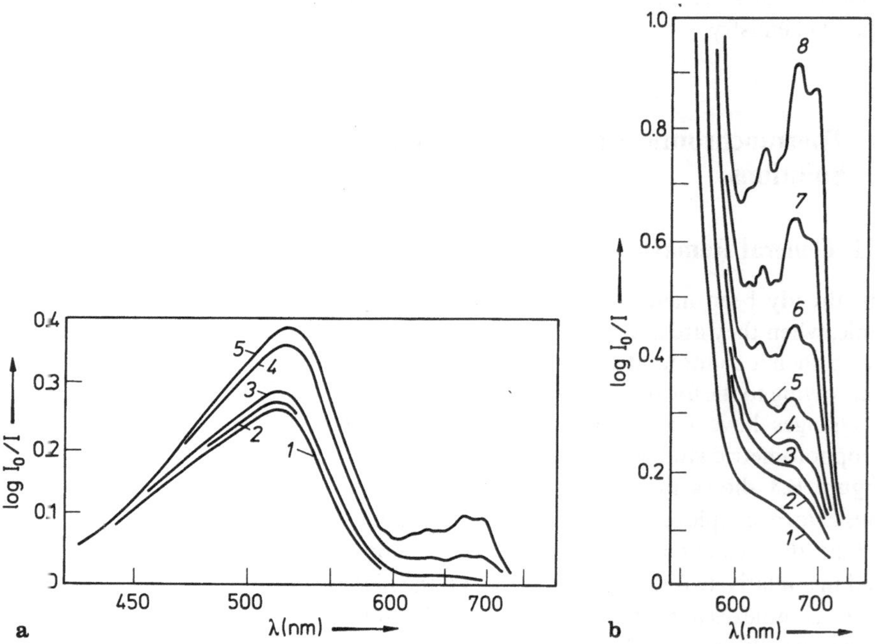

Fig. B-9 a, b. Absorption curves of solutions containing 0.5 M $CoCl_2$ and various amounts of $NaClO_4$. Curves *1–5* in **a** correspond to the solutions containing 0, 1, 3, 6, and 7.4 M $NaClO_4$, and curves *1–8* in **b** correspond to solutions with 0, 5.3, 5.5, 6.0, 6.4, 6.7, 7.0, and 7.3 M $NaClO_4$, respectively. Cell thickness: 1 mm in **a**, 1 cm in **b**; temp. 24–25 °C. (Taken from Mizutani and Sone [38])

in the corresponding $NiCl_2$ and $NiBr_2$ solutions, because of the low stability of $[NiX_4]^{2-}$ [37c] (cf. Chaps. B.VII and C.V).

It is thus clear that solutions of Co^{2+} which are several molar in Cl^- ions are equilibrium mixtures of various blue and pink species. We can expect such solutions to be much more thermochromic than those without an excess of Cl^- ions. In fact this is true. However, even the latter solutions can be made remarkably thermochromic, when a large amount of an inert electrolyte, such as $NaClO_4$, is added.

This interesting phenomenon, called the "inert salt effect", was discovered by Mizutani and Sone [38] and can easily be observed in test tubes. A dilute aqueous solution of $CoCl_2$ is pink and only slightly thermochromic. When a large amount of $NaClO_4$ (1—7 M) is added, the color at room temperature becomes reddish violet. Figure B.9a shows the corresponding spectral change, and Fig. B.9b shows vertically enlarged curves for the sharply split band appearing between 600–700 nm in Fig. B.9a. Upon heating to 40 °C, this same band for a solution with 6 M $NaClO_4$ becomes nearly four times higher, yielding a bluish solution.

Now we can combine these results with the data in Fig. B.8. The slight shift of the original band of the pink $[Co(H_2O)_6]^{2+}$ cation at 510 nm to ca. 525 nm, and its notable amplification, clearly indicate the formation of $[Co(H_2O)_5Cl]^+$ by the following reaction (cf. Chap. B.II):

$$[Co(H_2O)_6]^{2+} + Cl^- \rightleftharpoons [Co(H_2O)_5Cl]^+ + H_2O. \qquad \text{(Eq. B.10)}$$

This change is not caused by the increase in Cl^- ions, but is caused by a decrease in water activity (or its coordination power) caused by the inert salt.

Although a quantitative estimate of such an effect is difficult to obtain, we can get an idea of it from the book by Beck [39]. Table B.1 shows the concentrations of $NaClO_4$ and H_2O in pure water and four solutions, and those of "free water molecules" calculated under the assumption that 1, 3, 6 and 10 water molecules are bound in the hydration spheres of the salt ions, respectively. We can see that, even when only an average of 3 H_2O is bound to a pair of Na^+ and ClO_4^- (this is certainly an underestimate), the "free water concentration" decreases drastically at salt concentration above 3 M. This will shift the equilibrium in (Eq. B.10) to the right-hand side, similar to an increase in Cl^- concentration.

Table B.1. The concentrations of $NaClO_4$(C_{NaClO_4}), water(C_W), and "free water"($C_W - nC_{NaClO_4}$, n = 1, 3, 6, 10) in various $NaClO_4$ solutions (After Beck [39])

		Free water concentration (M)			
C_{NaClO_4}	C_W	$C_W - C_{NaClO_4}$	$C_W - 3C_{NaClO_4}$	$C_W - 6C_{NaClO_4}$	$C_W - 10C_{NaClO_4}$
0	55.51	55.51	55.51	55.51	55.51
0.082	55.23	55.15	54.98	54.74	54.41
1.008	52.74	51.68	49.57	46.69	42.16
2.952	47.51	44.51	38.52	29.80	17.55
9.396	30.06	20.74	2.094	–	–

At higher concentrations of $NaClO_4$, higher chloro-complexes begin to form, as more and more water molecules are squeezed out of the coordination sphere by the same effect. A look at Fig. B.8 suggests that the di- and tri-chloro species will only be formed in limited amounts, so that the main blue species appearing at high $NaClO_4$ concentrations will be $[CoCl_4]^{2-}$. This can be confirmed by a comparison of the sharply split peaks and shoulders of the resulting band with those of $[CoCl_4]^{2-}$ (cf. Fig. B.2). The mutual coincidence of their positions and shapes is fantastic, if the overlap of the band with the tail of the octahedral band is taken into account. Thus, when the $NaClO_4$ concentration is greater than ca. 6 M, the following reaction can be monitored spectroscopically:

$$[Co(H_2O)_6]^{2+} + 4Cl^- \rightleftharpoons [CoCl_4]^{2-} + 6H_2O. \quad \text{(Eq. B.11)}$$

It was also found that the perchlorates of Mg^{2+}, Ca^{2+} and Sr^{2+} all show the same kind of spectral effects, but much more strongly than $NaClO_4$ ($Mg^{2+} > Ca^{2+} > Sr^{2+}$). This is probably due to the much higher hydration energies of these divalent ions, which increase in this order.

It must be added that there are numerous similarities between concentrated salt solutions, or melts of salt hydrates formed at relatively low temperatures, and solutions in organic solvents. The dielectric constants of such solutions are much lower than that of pure water. For a salt concentration less than 2 M, the relation $D = D_0 + 2\bar{\delta}c$ holds in many cases, where D_0 and D are the dielectric constants of pure water and a salt solution, respectively, c is the concentration of the latter (in M), and the values of $2\bar{\delta}$ are given in Table B.2 [40]. Thus, D is ca. 56 in a 2 M solution of NaCl, while that of $MgCl_2$ is still much smaller. In this respect, there is also a notable similarity between such solutions and very hot water, in which such a decrease in D takes place (cf. Chap. B.I).

Table B.2. The values of $2\bar{\delta}$ for various salts (After Robinson and Stokes [40])

LiCl	−14	NaF	−12	KI	−16
NaCl	−11	KF	−13	$MgCl_2$	−30
KCl	−10	NaI	−15	$BaCl_2$	−28

Although notable thermochromic changes of $CoCl_2$ in pure water only take place at high temperatures and concentrations which are inconvenient to study, it is probable that the mode of the observed spectral changes will be similar to that observed in inert salt solutions of high concentrations. In both cases, the change takes place in water of diminishing polarity and coordination power. In fact this was found to be the case; Moriyama [41] carefully studied the thermochromism of $CoCl_2$ in pure water from 0 °C to 70 °C and found that only a slight and gradual deformation of the band at 510 nm takes place upon heating a 0.25 M solution, which is similar to the curves 1–3 of Fig. B.9a. In a 2 M solution, this deformation proceeds even further, while a group of peaks similar to those in curves 4–8 of Fig. B.9b rapidly grows at 600–700 nm. The general features of the observed spectral changes are similar to those depicted in these Figures. We can probably say that the changes which occur in concentrated aqueous $CoCl_2$ solutions above 100 °C can be observed in dilute solutions below 100 °C by adding a large amount of an inert salt.

In hot solutions, thermal motion hinders the interaction of water molecules with one another and with the solute. On the other hand, ionic solvation in concentrated inert salt solutions hinders the free orientation of water molecules in an electric field (leading to lower D) and their free interaction with the solute. The two situations are not the same, but similar in that the solvent-solute interactions are strongly weakened. In other words, the solvent in such solutions does not take much care of the solute, so that the latter gets quite "dry" even in water.

It is also interesting to arrange the modes of thermochromic changes in water, methanol and ethanol (and higher primary alcohols) in the following sequence, in which the predominant species at room temperature are boldfaced:

$$H_2O(+NaClO_4)\text{: } \mathbf{[Co(H_2O)_6]^{2+}} \rightarrow [Co(H_2O)_5Cl]^+ \rightarrow [CoCl_4]^{2-} \quad \text{(Eq.B.12a)}$$

$$MeOH\text{: } [Co(MeOH)_6]^{2+} \rightarrow \mathbf{[Co(MeOH)_5Cl]^+} \rightarrow [Co(MeOH)Cl_3] \quad \text{(Eq.B.12b)}$$

$$EtOH\text{: } ([Co(EtOH)_6]^{2+}/[Co(EtOH)_5Cl]^+) \rightarrow \mathbf{[Co(EtOH)_2Cl_2]} \quad \text{(Eq.B.12c)}$$

$$\text{Temp.: Low} \longleftrightarrow \text{High}$$

As one can easily see, that the temperature of the main (pink → blue) change decreases consistently in the order of $H_2O > MeOH > EtOH$. The main blue species that are formed in them are of $[CoCl_4]^{2-}$, $[Co(MeOH)Cl_3]^-$ and $[Co(EtOH)_2Cl_2]$, respectively. This is in agreement with the earlier statement made by Libuś, i.e., "the weaker the coordination of the solvent, the less ligands (i.e., anions) are needed to enter the coordination sphere of the metal ion and bring about a lowering of the coordination number from 6 to 4" [19].

It should be noted that this statement only refers to the predominant species. In fact it is rather difficult to answer the question: "How many Cl^- ions must enter the coordination sphere of $[Co(Solv)_xCl_y]^{(2-y)+}$, in order to change the coordination number (C.N. = x + y) from 6 to 4?"

In general, when y = 3 or 4, the complex is blue (C.N. = 4); on the other hand, when y = 0 or 1, the complex is pink or reddish violet (C.N. = 6) if the solvent molecule is not too big (cf. Chap. B.VI). In most cases, the change begins at y = 2. In most alcohols and acetone, the species formed at this point is one with C.N. 4 ($[Co(Solv)_2Cl_2]$), but according to Sawada and Suzuki [49] that in acetic acid has C.N. = 6, ($[Co(Solv)_4Cl_2]$), while an equilibrium between these two species is established with pyridine derivatives.

The C.N. of the dichloro species formed in aqueous solutions (cf. above) is also somewhat dubious; although Zeltmann et al. assumed it to be 4 [37], we suspect it to be 6. One of our reasons is that $CoCl_2 \cdot 6H_2O$ is composed of trans-$[Co(H_2O)_4Cl_2]$ and $2H_2O$; thus, it seems natural to assume the existence of the same species in solution, from which such crystals separate out upon concentration.

However, these are only minor problems which we do not need to worry about too much. We still have many more color phenomena of interest to discuss in this chapter.

B.IV.2 Thermochromism of Cobalt(II) Chloride in Reversed Micelles

Another interesting thermochromic phenomenon shown by aqueous $CoCl_2$ solutions is that of "reversed micelle sols" observed by Sunamoto and Hamada [42]. They prepared such sols by mechanically shaking a mixture of chloroform, an anionic or cationic surfactant, and a very small amount of a concentrated aqueous solution of $CoCl_2$ (3 M). A sketch of a micelle in the resultant colored sol is shown in Fig. B.10.

In this sol, the water is solubilized in chloroform as colloidal particles, which are stabilized by a surrounding layer of surfactant ions with their polar ends protruding into the water. Since the situation is the reverse of that found in the micelles of common soap solutions, the sol is called a reversed micelle sol.

The micelle contains $CoCl_2$ and the counter ions of the surfactant. Figure B.10 illustrates the situation when the detergent used is N,N-dimethyl-N,N-dioctadecylammonium chloride (QD). A simple calculation using their data indicates that the sol

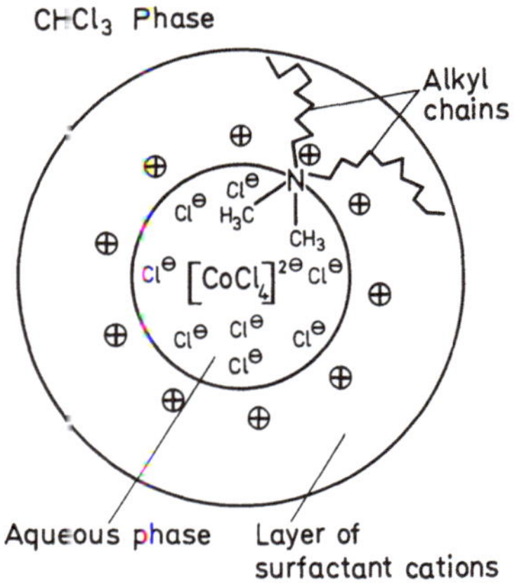

Fig. B-10. Sketch of a reversed micelle formed by QD in $CHCl_3$ and the aqueous $CoCl_2$ solution solubilized in it

contains ca. 2×10^{-1} M of the surfactant, 3×10^{-3} M $CoCl_2$ and 5.5×10^{-2} M H_2O, so that a micelle will contain these components in approximately the same proportions. The sol is deep blue and its spectrum is similar to that of $[CoCl_4]^{2-}$ (cf. Fig. B.1). This is because (i) the concentration of $CoCl_2$ in the micelle is high, (ii) there is a large excess of Cl^- ions in the same micelle originating from the surfactant and (iii) nearly all of the water in it is interacting with the positive ends of the surfactant, so that the activity of the water in the micelle is much lower than in common water.

A similar blue sol is formed when hexadecyltrimethylammonium bromide (CTAB) is used instead of QD. In this case, the band of the blue cobalt(II) species is notably shifted towards the red (ca. 20 nm), indicating that $[CoBr_4]^{2-}$ and/or mixed complex species, such as $[CoCl_2Br_2]^{2-}$, are formed in the micelle.

It is interesting to note that the water added to such sols (followed by vigorous shaking) is absorbed into the reversed micelles, thereby diluting the contents and changing the color into the pale pink of $[Co(H_2O)_6]^{2+}$. If, for example, a CTAB-containing sol is treated in this way and the intensity of the band of the blue species is plotted against the concentration of water, a sigmoid curve shown in Fig. B.11a is obtained. The shape of this curve indicates that the added water at first interacts with the polar ends of the surfactant; the concentration of the free water increases only after the polar ends of the inner wall of the micelle have been saturated with water.

The pink solutions obtained in this manner are remarkably thermochromic. Figure B.11b shows that the same band which disappeared upon the addition of water appears again with only slight heating to ca. 50 °C, forming a curve which is nearly a mirror image of Fig. 11a.

Using these and two other surfactants, Sunamoto and Hamada tried to analyze the spectral data to obtain more quantitative information on the equilibria involved. We shall not discuss these results in detail, but try to consider about the interaction between the water molecules in the micelle and the polar ends of the surfactant.

As can be seen from the composition of the sol, the amount of water in a micelle is much smaller than the amount of surfactant surrounding it. It is by no means enough to combine with all of the polar ends of the surfactant, and the accompanying halide ions. Therefore, both components of the surfactant are only very weakly hydrated and a considerable percentage probably forms loose ion pairs. If $[Co(H_2O)_6]^{2+}$ is introduced

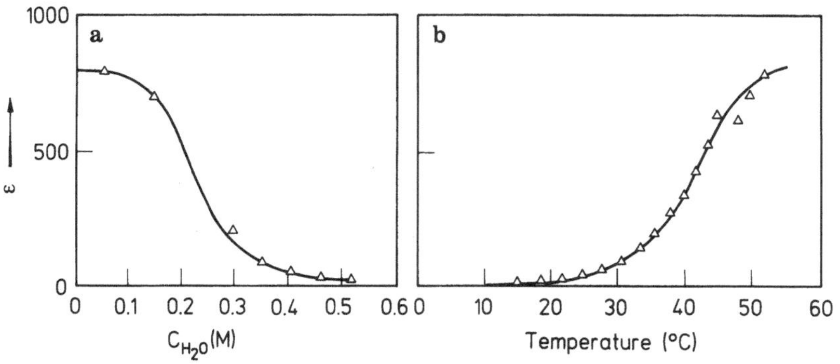

Fig. B-11 a, b. The effects of water **(a)** and temperature **(b)** on the intensity of the band at 700 nm of $CoCl_2$ solubilized in CTAB-containing reversed micelles. (Taken from Sunamoto and Hamada [42])

into such a micelle, its water molecules will be readily squeezed out of the coordination sphere, for the sake of better hydration of the ambient particles which are in large excess. The remaining Co^{2+} is then attacked by the poorly solvated, reactive halide ions to form blue anions.

With the addition of more water, all of the particles in the micelle are increasingly hydrated, but, as mentioned above, the water is at first consumed to improve the hydration of the predominant surfactant particles. Only after the latter have been satisfied to a certain extent, do the blue anions dissociate into $[Co(H_2O)_6]^{2+}$ and X^-. However, even when the sol has just become pink, the water molecules surrounding Co^{2+} are in a very marginal state, since there are only about 2 moles of water per mole of surfactant (cf. Fig. B.11). Even a small amount of thermal energy is thus sufficient to cause the "squeezing" again, making the solution once more blue.

In their discussions, Sunamoto and Hanada assumed that the octahedral species is $[Co(H_2O)_4X_2]$. This, however, does not matter so much for the present discussion.

The differences between the various kinds of water molecules in a micelle, i.e., those bound to the polar end of the surfactant and those surrounding X^- and Co^{2+}, will by no means be distinct. So it may be better to say that all the water molecules in a micelle are under the combined influence of the charged particles in the system. The same will also apply to the solutions in concentrated inert salts described in the last section. Hence, the idea implied in Table B.1 should only be taken as an oversimplified visualization.

At any rate, concentrated electrolyte solutions are another "twilight zone" in solution chemistry, similar to superheated solutions (cf. Chap. B.I), which merit further study in the future.

Sunamoto and collaborators also studied the sols containing halide complexes of Ni^{2+} and Cu^{2+} (cf. Chap. B.VII) in reversed micelles, and applied the results in order to mimic the active sites of metalloenzymes [42a, 42b].

B.V Chromotropic Phenomena of Cobalt(II) Chloride in Aprotic Solvents

So far, our discussions have been confined to aqueous and alcoholic solutions of $CoCl_2$. What is the situation in an aprotic solvent, such as acetone, DMF, acetonitrile etc.?

In general, blue solutions are formed in most of these solvents, which show the characteristic "tetrahedral" spectra. No drastic change in color is observed, when they are heated, cooled, or compressed. In certain solvents such as acetone or DMSO, much more water must be added to the solutions to make them pink than in the case of the alcoholic solutions. It seems, therefore, that the blue species formed in them are quite stable towards chromotropic changes.

Spectral observations, however, have shown that the shape of their tetrahedral band changes remarkably with temperature and, in certain cases (e.g. in acetone), at high pressures. Examples of the thermochromic effect and the piezochromic effect, taken from the works of Moriyama [41] and Ishihara et al. [34], respectively, are shown in Figs. B.12 and B.13.

Although these solutions have not been as thoroughly studied as in the case of the aqueous and alcoholic solutions, these data strongly indicate that the equilibrium of

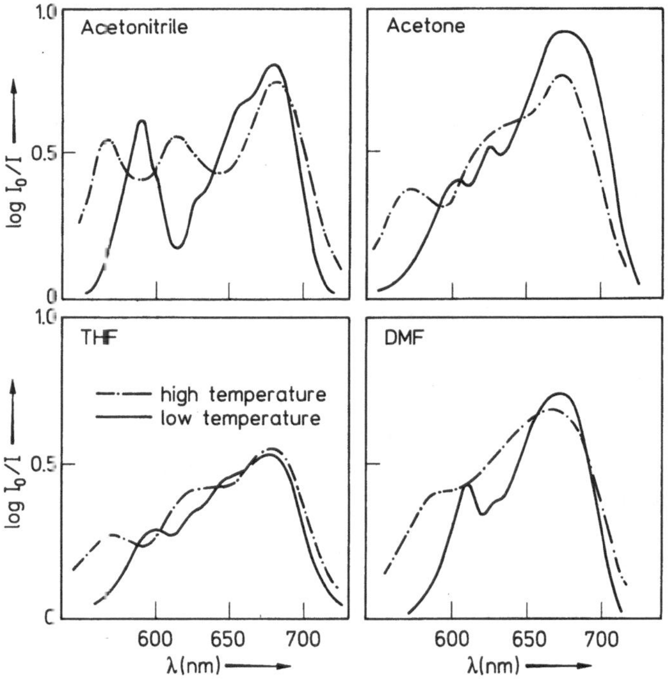

Fig. B-12. Sketch of thermochromic spectral changes of $CoCl_2$ in four aprotic solvents. The "high" and "low" temps. are: 70 and $-30\,°C$ for acetonitrile, 55 and $-60\,°C$ for acetone, 60 and $-80\,°C$ for THF, and 90 and $-35\,°C$ for DMF. Conc.: $(4-5) \times 10^{-3}$ M, cell thickness: 5 mm. These curves were drawn without corrections for the volume changes of the solutions with temperature; in doing so, the low-temp. and high-temp. curves sink and rise somewhat, respectively. (Taken from Moriyama [41])

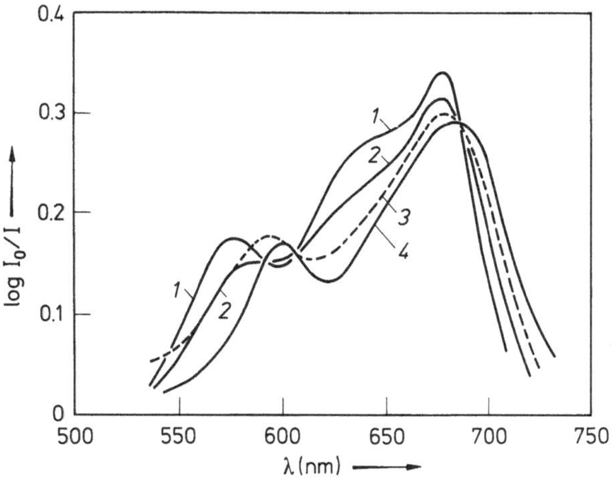

Fig. B-13. Absorption spectra of $CoCl_2$ in acetone under various pressures (P). Conc.: 2.05 $\times 10^{-3}$ M; cell thickness: 7.0 mm. Curves *1* to *4* correspond to P of 1, 2×10^3, 4×10^3, and 8×10^3 kg cm^{-2}, respectively. (Taken from Ishihara et al. [34])

the type of (Eq. B.6), which was proposed in connection with the thermochromism in sec-butanol, is of greater importance in this case. This equilibrium is associated with an ionization of the type given in (Eq. B.4), so that the thermochromic equilibria in these solvents take on the character of a disproportionation reaction, such as:

$$2[Co(Solv)_2Cl_2] + 2Solv \underset{\text{Heat}}{\overset{\text{Cool}}{\rightleftharpoons}} [Co(Solv)_5Cl]^+ + [Co(Solv)Cl_3]^-$$

Blue Pink Blue

(Eq. B.13a)

or

$$3[Co(Solv)_2Cl_2] + 2Solv \underset{\text{Heat}}{\overset{\text{Cool}}{\rightleftharpoons}} [Co(Solv)_6]^{2+} + 2[Co(Solv)Cl_3]^-$$

Blue Pink Blue

(Eq. B.13b)

As the intensity of the absorption band of $[Co(Solv)Cl_3]^-$ is generally known to be much higher (e. g., ca. 50% in the case of acetone [35]) than that of $[Co(Solv)_2Cl_2]$, the apparent intensity of the blue color of the solution is expected to change only a little by such reactions. While there are still some ambiguities in the explanation of the spectra in acetone, the data in the other three solvents in Fig. B.12 are reasonably compatible with this point of view. The piezochromism in acetone in Fig. B.13 can be explained well using the equilibrium shown in (Eq. B.13b). In this case the effect of compression is in the same direction as cooling, as in the case of alcoholic solutions.

It may be added that the existence of such disproportionation or "autocomplexation equilibria" in these and other aprotic solvents at room temperature has already been pointed out by a number of investigators. This can be seen, e.g., from the distribution charts of various complex species obtained in the spectral studies of Libuś (acetronitrile) [18] and Sawada (acetone) [35]. Buffagni and Dunn [20] even stated that the disproportionation goes still further in DMF, producing small but noticeable amounts of $[CoCl_4]^{2-}$.

On the other hand, there is also structural reason for such equilibria in these solutions, in comparison with the simple ionization equilibria (Eq. B.4) or (Eq. B.5a–b) observed in primary alcohols. It was pointed out earlier (B.III) that the weaker solvation of the Cl^- ion by secondary or tertiary alcohols is expected to be a driving force that prevents the ionization of $[Co(ROH)_2Cl_2]$ and helps to convert some of the latter into $[Co(ROH)Cl_3]^-$. It is very likely that this same factor is operating much more effectively in aprotic solvents, leading to the equilibria given in (Eqs. B.13a–b) and making the solutions thermo- and piezochromic. For example, let us compare iso-PrOH and acetone. Both of these solvents contain polar molecules in which the negative charge is concentrated on the O atom. In iso-PrOH, much of the positive charge is concentrated on the H atom of the OH group, while the rest is smeared over the isopropyl group. In acetone, however, the entire positive charge is smeared over the $(CH_3)_2C$-group. There is not a single atom on which the positive charge is mainly concentrated.

Due to this charge distribution, acetone is expected to be much less effective in solvating a negative ion than iso-PrOH, which can easily do so by H-bonding. It can naturally solvate a cation with ease, but can only weakly bind an anion because of its smeared-out positive charge.

In fact, the values of AN and DN for acetone are 12.5 and 17.0, respectively. Compared with those of the alcohols cited above (Chap. B.II) the large decrease in AN is quite evident.

The Cl^- ions formed in acetone by the ionization reaction,

$$[Co(acetone)_2Cl_2] + (3\text{–}4)acetone \rightleftharpoons ([Co(acetone)_5Cl]^+ \text{ or } [Co(acetone)_6]^{2+}) + (1\text{–}2)Cl^-, \qquad \text{(Eq. B.14)}$$

are thus unstable and reactive, so that they tend to combine with the remaining $[Co(acetone)_2Cl_2]$ by driving out an acetone molecule:

$$[Co(acetone)_2Cl_2] + Cl^- \rightleftharpoons [Co(acetone)Cl_3]^- + acetone. \qquad \text{(Eq. B.15)}$$

The negative charge in the resulting complex is smeared over a number of atoms, so it can be formed even in a poorly anion-solvating medium like acetone. If this second reaction goes to completion, the net change in solution will be approximated by a reaction of the type given in (Eqs. B.13a–b).

The same situation is expected to occur in most of the polar aprotic solvents, in which the negative charge is localized on a certain single atom (N,O or sometimes S) but the positive charge is delocalized over the rest of the molecule. These structural features, therefore, seem to be the main reason for the importance of the equilibria of the types (Eqs. B.13a–b) in the chromotropic phenomena in aprotic solvents.

In terms of the donor-acceptor concept to be described later (cf. Chaps. C and D), such a molecule is a strong donor, but a weak acceptor. This can be seen, for example, by comparing the AN and DN values of these solvents with those of alcohols and water, which are strong donors and strong acceptors (cf. Table C.4). For a more direct comparison of the solvation energies, one can also consult the work of Ahrland [43], in which the enthalpy of transfer, ΔH_{tr} (W $\rightarrow$ S), between water and organic solvents is tabulated for various cations and anions. Although complications sometimes do occur, the general tendencies of the changes in these and related values are in line with our point of view.

B.VI Chromotropic Phenomena of Other Cobalt(II) Salts in Aqueous and Organic Solvents

The reader may feel that we have discussed $CoCl_2$ solutions too extensively. However, this is because the colors of these solutions have been the object of innumerable investigators since the discovery of $CoCl_2$ itself, and far more information is available on $CoCl_2$ than on any other simple salt. Although the explanations proposed for the color changes are, in most cases, only qualitative and imcomplete at present, a number of important problems encountered in modern solution chemistry are involved; therefore, their discussion will also serve as a useful introduction to the phenomena to be described later.

The bromide and iodide of cobalt(II) also show color changes similar to those of $CoCl_2$ in solution. Such changes were very often studied together with those of $CoCl_2$. Tetrahedral $[CoBr_4]^{2-}$ and $[CoI_4]^{2-}$ are formed in aqueous solutions of high halide concentration. Various complexes containing Br^- or I^- and solvent molecules are formed in organic solvents, in much the same way as with $CoCl_2$. The low-temperature thermochromism of $CoBr_2$ in alcohols has been studied by Kobayashi [44]. Its piezochromism in acetone was studied by Ishihara et al. [34]. The general characteristics of the observed spectral changes, including the notable difference between primary and secondary alcohols, are quite similar to those of $CoCl_2$. However, as is known with other transition metal ions of the first row, Co^{2+} forms complexes with Br^- that are somewhat less stable than those with Cl^-, so that the blue species become notably hard to form. For example, higher temperatures are needed to convert the pink complexes $[Co(Alcohol)_6]^{2+}$ or $[Co(Alcohol)_5Br]^+$ into the blue complex $[Co(Alcohol)_2Br_2]$, and smaller pressures are sufficient to convert the latter into the former [32].

Oxo-acid salts of cobalt(II), such as the nitrate, perchlorate, or sulfate, also exhibit thermochromic phenomena. The color changes of the nitrate and sulfate in superheated aqueous solutions have already been mentioned in Chap. B.I. The phenomena recently reported by Soukup [45, 46] are especially simple and interesting. He compared the colors of $Co(ClO_4)_2.6H_2O$ or $Co(NO_3)_2.6H_2O$ in HMPA, pyridine, DMSO, formamide and DMF.

All of these solvents are known to be very strong donors. Their donor numbers (DN; cf. Chap. C) are much higher than that of water. The perchlorate or nitrate ions, on the other hand, are known to form no complex or only weak ones in water, respectively. Therefore, Soukup expected the water of hydration and the anions of these salts to be driven off by dissolving the salts in one of these solvents. The Co^{2+} in solution will then assume an octahedral $[Co(Solv)_6]^{2+}$ (pink) or tetrahedral $[Co(Solv)_4]^{2+}$ (blue) state, since complexes of other stereochemistries are rare with the Co^{2+} ion.

Soukup found that the HMPA solution is blue but not thermochromic. He concluded that it contains Co^{2+} in the form of $[Co(HMPA)_4]^{2+}$ which is stable even at higher temperatures. On the other hand, the pyridine solution is pink. The other three solutions are thermochromic, being reddish when cold and bluish when hot. This observation indicates the existence of the simple equilibrium:

$$[Co(Solv)_6]^{2+} \rightleftharpoons [Co(Solv)_4]^{2+} + 2Solv \qquad \text{(Eq. B.16)}$$

Soukup ascribed this change to a simple steric effect. Comparing the formulae of the ligands concerned we can readily understand the situation. Apparently, the bulkiness of the non-coordinating groups and their strong mutual repulsion in the coordination sphere will result in a smaller C.N. for the HMPA complex. The less bulky ligands, DMSO, formamide and DMF, form complexes which will tend to be octahedral at low temperature and tetrahedral at higher temperatures, since thermal agitation reinforces the steric effect. Finally the complex becomes 6-coordinate with py, which is the least bulky of the ligands used.

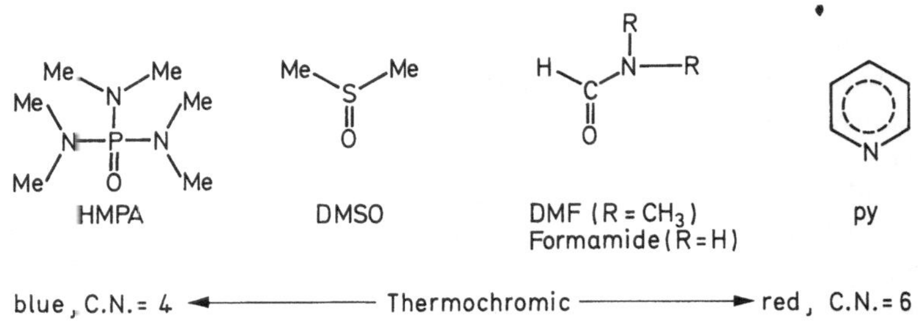

In this connection, the work of Abe [47] on the spectra and conductivities of various bivalent metal perchlorates in HMPA is of interest. She concluded that a 4-coordinate tetrahedral species $[M(HMPA)_4]^{2+}$ exists in solution when the metal ion (M^{2-}) is Mn^{2+}, Fe^{2+}, Co^{2+} or Cu^{2+}. However, with Ni^{2+} the resulting solution is strongly thermochromic as shown in Fig. B.14.

At 60–70 °C, the solution shows two strong peaks between 550 and 650 nm, which are quite similar to those shown by $[NiCl_4]^{2-}$ (cf. Chap. B.VII), and thus characteristic of the tetrahedral species $[Ni(HMPA)_4]^{2+}$. The same peaks are also observed in the spectrum of the solid $[Ni(HMPA)_4](ClO_4)_2$ complex. The intensities of these peaks decrease rapidly upon cooling, and a weaker band appears at ca. 470 nm. Upon further cooling the main peaks of $[Ni(HMPA)_4]^{2+}$ continue to decrease in intensity. The band at 470 nm is also gradually weakened and shifted towards shorter wavelengths.

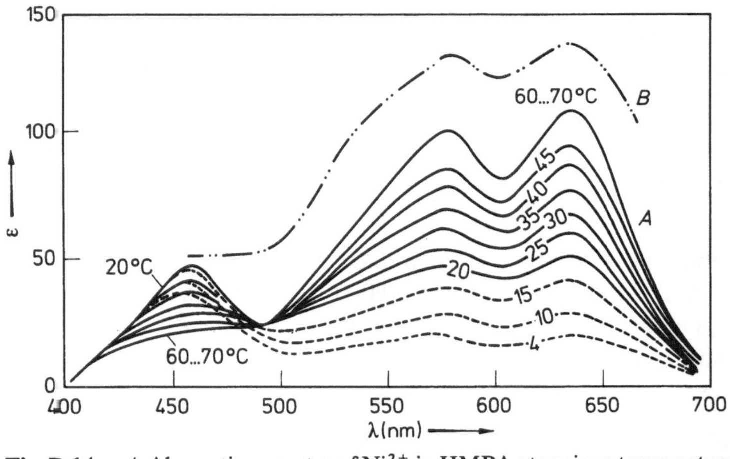

Fig. B-14. *A*: Absorption spectra of Ni^{2+} in HMPA at various temperatures. Conc. = 4.1×10^{-3} M. *B*: Solid reflection spectrum of $[Ni(HMPA)_4]$ $(ClO_4)_2$, arbitrary scale for ε. (Taken from Abe and Wada [47])

From these data, Abe assumed the following thermochromic equilibria:

$$\underset{\substack{\text{tetrahedral} \\ \text{I}}}{[Ni(HMPA)_4]^{2+}} \rightleftharpoons \underset{\substack{\text{square planar} \\ \text{II}}}{[Ni(HMPA)_4]^{2+}} \rightleftharpoons \underset{\substack{\text{octahedral} \\ \text{III}}}{[Ni(HMPA)_6]^{2+}} \qquad \text{(Eq. B.17)}$$

high ← temperature → low

She ascribed the band at 470 nm to another $[Ni(HMPA)_4]^{2+}$ complex with a square planar structure (II) and its shift at lower temperatures to the formation of an octahedral complex (III) with much weaker absorption bands (for the reasoning behind these assignments, cf. Chap. C). Since the $3d^8$ configuration of Ni^{2+} favors the formation of II or III but not I (cf. Chaps. C.II and C.V), it is conceivable that Ni^{2+} will tend to attract as many ligand molecules as possible to form III at very low temperatures. The latter complex will be too highly crowded in its coordination sphere, so that it will easily lose two ligands to form II at slightly higher temperatures. Upon further heating, II is converted into the tetrahedral modification I of higher energy. In this case it seems that the thermochromism of Co^{2+} in DMSO, DMF etc. observed by Soukup is appearing in a more complicated way.

Further information of special interest in this field can be obtained from the recent thermodynamic studies of Sawada [35, 48–50], covering the chloro- and bromo-complex formation of Co(II) in acetone, acetic acid and pyridine, as well as the NMR studies of Dickert and Hellmann [51] on the solvation of Co^{2+} in organic solvent mixtures. The works of Kojima et al. [52–53], who analyzed the thermochromic and piezochromic data of $CoCl_2$, $CoBr_2$ and CoI_2 in pyridine on a reasonably simplified basis, are also of interest.

B.VII Chromotropic Phenomena of Copper(II) and Nickel(II) Halides in Solution

A number of colored salts of the first-row transition metals other than those of cobalt(II) exhibit chromotropic changes in solution. Most of them are not as dramatic as those of the cobalt(II) salts. The thermochromism of the halides of copper(II) and nickel(II), which was already referred to in Chap. B.I, shall now be considered in more detail.

It is well known that the color of an aqueous solution of copper(II) chloride changes reversibly from blue to green, when heated. If the concentration and temperature are high enough, the solution assumes a yellowish or brownish color. This change in color shows many similarities with that of $CoCl_2$: it occurs at lower temperatures with the addition of Cl^- ions (with a large excess of them, the solution is green even at room temperature); with the addition of organic solvents (the resulting solutions are also often green), and even with the addition of large amounts of $NaClO_4$ [38] to an aqueous solution.

The spectral changes accompanying such changes in color are not as attractive as in the case of $CoCl_2$. They consist of a gradual shift of the absorption band of $[Cu(H_2O)_4]^{2+}$ (ca. 800 nm) towards the infrared region with an accompanying increase in intensity, and the appearance of a strong absorption band in the near ultraviolet, which also gradually shifts into the visible region (cf. Fig. B.16). These results show that the ligand field strength in the $[Cu(H_2O)_4]^{2+}$ ion is weakened, and a CT band (or a series of such bands) of the ($Cl^- \rightarrow Cu^{2+}$) type appears, as the H_2O is successively substitued by Cl^- ions. Thus, it seems that this color change is due to a shift in the equilibria:

$$[Cu(H_2O)_4]^{2+} \rightleftharpoons [Cu(H_2O)_3Cl]^+ \rightleftharpoons [Cu(H_2O)_2Cl_2] \rightleftharpoons [Cu(H_2O)Cl_3]^- \rightleftharpoons [CuCl_4]^{2-} \qquad \text{(Eq. B.18)}$$

or similar equilibria involving organic solvent molecules.

Copper(II) bromide is also thermochromic in solution. In this case the situation is more complicated. When higher bromo-complexes ($[Cu(Solv)Br_3]^-$ or $[CuBr_4]^{2-}$) are formed, the CT band appears in the middle of the visible region (around 600 nm). The resulting solutions are purple-red. Moreover, redox equilibria leading to Cu^+ and Br_2/Br_3^- take place in certain solvents (e.g. acetonitrile) [54]. Inspite of such troubles, these solutions were studied by many investigators, particularly with reference to the photochromism caused by strong illumination to study their suitability for use in light-protecting goggles. References can be found in the reviews of Day (cf. Ref. [2] and [3] in Chap. A).

Nickel(II) chloride is much less thermochromic than copper(II) chloride. The results of Griffiths and Scarrow are of special interest [55]. An aqueous solution of $NiCl_2$ is green and exhibits three weak bands at ca. 395, 725, and 1180 nm, due to $[Ni(H_2O)_6]^{2+}$ (cf. Chap. C.I). When a large amount of $[N(CH_3)_4]Cl$(5 M) is added to a dilute solution of $Ni(ClO_4)_2.6H_2O(1 \times 10^{-3}M)$, the color and spectrum of $[Ni(H_2O)_6]^{2+}$ are changed only a little. Heated to 110–120 °C, the solution assumes a deep blue color. Figure B.15 shows the spectral changes involved.

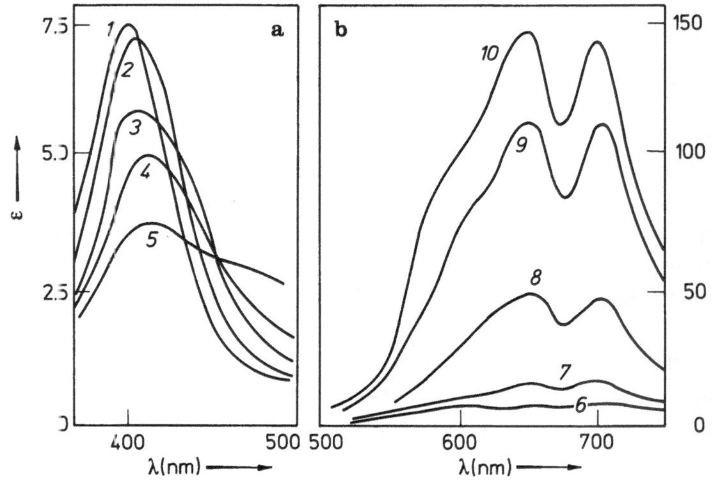

Fig. B-15 a, b. Thermochromism of a concentrated aqueous solution (5.6 M) of $[(CH_3)_4N]Cl$ containing ca. 1×10^{-3} M of $Ni(ClO_4)_2$. **(a):** Deformation of the band of octahedral species at ca. 400 nm, temp. (from *1* to *5*): 24, 72, 93, 104, and 114 °C. **(b):** Appearance of the peaks of tetrahedral species. Temp. (from *6* to *10*): 20, 40, 60, 83, and 110 °C. (After Griffiths and Scarrow [55])

The strong peaks at 654 and 705 nm, which appear in a heated solution, are quite similar to those observed in salt melts containing $NiCl_2$ (e.g. those in CsCl-$ZnCl_2$) [56] or in organic solutions of $NiCl_2$ containing a large excess of Cl^- (e.g. in acetonitrile or trimethyl phosphate) [22, 49]. The formation of tetrahedral $[NiCl_4]^{2-}$ was indicated in both cases. Therefore, in the presence of a large amount of $[N(CH_3)_4]Cl$, this same complex can also be formed in aqueous solutions at higher temperatures.

In contrast to $[CoCl_4]^{2-}$, which is readily formed in aqueous solutions of high Cl^- concentration, $[NiCl_4]^{2-}$ is much less stable. Even hot solutions of Ni^{2+} containing large amounts of HCl, LiCl, NaCl or $[N(C_2H_5)_4]Cl$ do not show its blue color. With Br^- or I^-, aqueous thermochromic solutions containing $[NiBr_4]^{2-}$ or $[NiI_4]^{2-}$ can not be obtained using any cation.

However, solutions containing $[NiBr_4]^{2-}$ or $[Ni\,I_4]^{2-}$ can be obtained by adding a large excess of Br^- or I^- in organic solvents. These complexes exhibit spectra similar to that of $[NiCl_4]^{2-}$. The peaks are shifted towards longer wavelengths in the order $Cl^- \rightarrow Br^- \rightarrow I^-$ [58].

Griffiths and Scarrow assumed that the small spectral change observed at ca. 400 nm is due to the formation of $[Ni(H_2O)_5Cl]^+$, and determined values of K and ΔH for the equilibrium:

$$\underset{\text{green}}{[Ni(H_2O)_5Cl]^+} + 3Cl^- \rightleftharpoons \underset{\text{blue}}{[NiCl_4]^{2-}} + 5H_2O \qquad \text{(Eq. B.19)}$$

at various temperatures from their spectral data. At 89 °C the "green to blue" ratio is approximately 1:1. It changes to about 1:50 at 120 °C.

The role of the $[N(CH_3)_4]^+$ ion in this thermochromism is peculiar. According to Griffiths and Scarrow, this cation can be taken to be a very small particle in water. Although one would intuitively think Li^+ or Na^+ are smaller, they are heavily hydrated in aqueous solutions and will behave like much larger particles. $[N(CH_3)_4]^+$ is a quasi-spherical ion, exhibiting paraffin-like surface groups. It cannot be hydrated via hydrogen-bond formation as in the case of NH_4^+. So in an aqueous solution it hardly disturbs the water structure. Since such unhydrated small cations are always present around each Cl^- ion (the mean anion-cation distance in a 5 M solution of $[N(CH_3)_4]Cl$ is only 0.6 nm), the hydration of the Cl^- ions is notably weakened, so that they get more "naked" and reactive than the Cl^- ions in other solutions. The otherwise impossible formation of $[NiCl_4]^{2-}$ in aqueous solutions is thus realized in solutions containing $[N(CH_3)_4]Cl$.

This model of Griffiths and Scarrow is very interesting, but, as those used in the case of color changes with cobalt(II) chloride, does not seem to completely answer all questions. Griffiths and Scarrow also reported spectral data for the thermochromic solutions of Ni^{2+} and a large excess of Cl^- ions in ethylene glycol and glycerol [59]. Due to the high boiling points, they could follow the drastic changes in color to nearly 200 °C.

Angell and Gruen [60] studied the spectra of Cu^{2+} and Ni^{2+} in very concentrated and hot solutions of $MgCl_2$. Some of the solutions contained nearly or less than 6 moles of H_2O per $MgCl_2$. They were, in fact, melts of the composition $MgCl_2 \cdot nH_2O$. Using special high-pressure cells, they heated such solutions up to ca. 320 °C. Some of their results are shown in Fig. B.16.

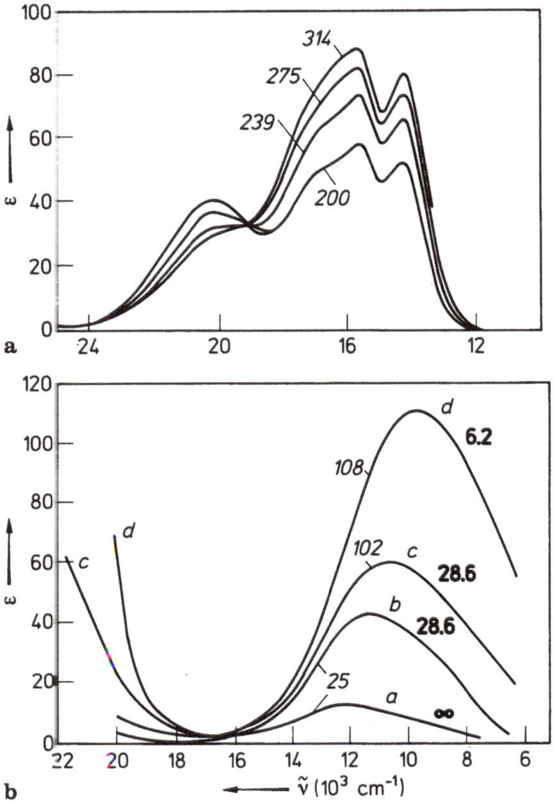

Fig. B-16 a, b. Thermochromism of Ni^{2+} in a melt of the composition $MgCl_2 . 5H_2O$ **(a)**, and that of Cu^{2+} in solutions and melts of various compositions $MgCl_2$ nH_2O **(b)**. The numbers in italic are the temperatures (°C); those in boldface are n. (After Angell and Gruen [60])

Comparing these data with the spectra of related solutions, solids or melts, they concluded that there is an equilibrium for Ni^{2+},

$$[NiCl_6]^{4-} \rightleftharpoons [NiCl_4]^{2-} + 2Cl^-, \qquad \text{(Eq. B.20)}$$

while $[CuCl_4]^{2-}$ is formed through a series of ligand-exchange reactions (cf. Eq. B.18). These studies thus lead us to another horizon in inorganic thermochromism, at which molten salt chemistry and solution chemistry shake hands with each other.

References

1. Howell, O.R. et al.: Proc. Roy. Soc. **A 142**, 587 (1933)
2. Cotton, F.A., Soderberg, R.H.: Inorg. Chem. **3**, 1 (1964)
3. Katsurai, T., Sone, K.: Kolloid-Z. **163**, 70 (1959)
4. Katsurai, T., Sone, K.: Bull. Chem. Soc. Jpn. **41**, 520 (1968)
5. Katsurai, T., Makide, Y.: Sci. Pap. Inst. Phys. Chem. Res. Tokyo **68**, 100 (1974)
6. Katsurai, T.: ibid. **72**, 125, (1978)
7. Kortüm, G.: "Lehrbuch der Elektrochemie", 4. Aufl., Verlag Chemie, Weinheim (1966), p. 174
8. Lüdemann, H.D., Franck, E.U.: Ber. Bunsen-Ges. **71**, 455 (1967)
9. Franck, E.U.: "Plenary Lectures from XXIInd International Congress of Pure and Applied Chemistry, Sydney 1969", Butterworths, London (1970), p. 13

10. Giggenbach, W.: J. Inorg. Nucl. Chem. **30**, 3189 (1968)
11. Ellis, A. J., Giggenbach, W.: Geochim. et Cosmochim. Acta **35**, 247 (1971)
12. Giggenbach, W.: Inorg. Chem. **10**, 1306 and 1308 (1971)
13. Giggenbach, W.: J. Chem. Soc. Dalton **1973**, 729
14. Katzin, L., Gebert, E.: J. Am. Chem. Soc. **72**, 5464 (1950)
15. Cotton, F. A. et al.: ibid. **83**, 4690 (1961)
16. Libuś, W.: Roczniki Chem. **33**, 931 and 951 (1959)
17. Libuś, W. et al.: ibid. **34** 29 (1960)
18. Libuś, W.: "Proceedings of 7th International Conference on Coordination Chemistry", Stockholm-Uppsala (1962), p. 349
19. Libuś, W.: "Theory and Structure of Complex Compounds", Pergamon Press, Oxford (1964), p. 537
20. Buffagni, S., Dunn, T. M. J.: J. Chem. Soc. **1961**, 5105
21. Fine, D. A.: J. Am. Chem. Soc. **84**, 1139 (1962)
22. Baaz, M. et al.: Monatsh. Chem. **93**, 1416 (1963)
23. Gutmann, V., Wegleitner, K. H.: ibid. **99**, 368 (1968)
24. Gutmann, V., Bohunovsky, O.: ibid. **99**, 740 (1968)
25. Tschebull, W. et al.: Z. anorg. allg. Chem. **416**, 323 (1975)
26. Cotton, F. A., Wilkinson, G. W.: "Advanced Inorganic Chemistry", 3rd Ed., Interscience, New York (1972), p. 881
27. Nieuwpoort, W. C. et al.: Rec. Trav. Chim. Pays-Bas **85**, 397 (1966)
28. Sone, K. et al.: Monatsh. Chem. **107**, 271 (1976)
29. Smithson, J. M., Williams, R. T. P.: J. Chem. Soc. **1958**, 457
30. Osugi, J., Kitamura, Y.: Nippon Kagaku Zasshi **89**, 569 (1968)
31. Osugi, J., Kitamura, Y.: ibid. **90**, 640 (1969)
32. Osugi, J., Kitamura, Y.: ibid. **90**, 889 (1969)
33. Kitamura, Y.: Rev. Phys. Chem. Japan **39**, 1 (1969)
34. Ishihara, I. et al.: ibid. **44**, 11 (1974)
35. Sawada, K. et al.: J. Inorg. Nucl. Chem. **43**, 3263 (1981)
36. Kato, M. et al.: Z. phys. Chem., Neue Folge **35**, 348 (1962)
37. Zeltmann, A. H. et al.: J. Phys. Chem. **72**, 121 (1968)
37a. Bjerrum, J. et al.: Acta Chem. Scand. **A 29**, 326 (1975)
37b. Libuś, Z.: Polish J. Chem. **53**, 1971 (1979)
37c. Libuś, Z. et al.: Polish J. Chem. **52**, 793 (1978)
38. Mizutani, K., Sone, K.: Z. anorg. allg. Chem. **350**, 216 (1967)
39. Beck, M. T.: "Chemistry of Complex Equilibria", Van Nostrand, London (1970), p. 170
40. Robinson, R. A., Stokes, R. H.: "Electrolyte Solutions", Butterworths, London (1970), 2nd Ed., p. 18
41. Moriyama, K. et al. (Ochanomizu University): unpublished data (1974)
42. Sunamoto, J., Hamada, T.: Bull. Chem. Soc. Jpn. **51**, 3130 (1978)
42a. Sunamoto, J. et al.: Inorg. Chem. **19**, 3668 (1980)
42b. Sunamoto, J., Kondo, H.: Inorg. Chim. Acta **92**, 159 (1984)
43. Ahrland, S.: Pure & Appl. Chem. **54**, 1451 (1982)
44. Kobayashi, F. et al. (Ochanomizu University): unpublished data (1977)
45. Soukup, R. W.: Chemie in unserer Zeit **17**, 129 (1983)
46. Soukup, R. W.: ibid. **17**, 163 (1983)
47. Abe, Y., Wada, G.: Bull. Chem. Soc. Jpn. **53**, 3547 (1980)
48. Sawada, K. et al.: J. Inorg. Nucl. Chem. **42**, 1471 (1980)
49. Sawada, K., Suzuki, T.: ibid. **43**, 2301 (1981)
50. Sawada, K. et al.: J. Chem. Soc. Dalton **1983**, 1565
51. Dickert, F. L., Hellmann, S. W.: Ber. Bunsenges. Phys. Chem. **87**, 513 (1983)
52. Kojima, K. et al.: Bull. Chem. Soc. Jpn. **56**, 684 (1983)
53. Kojima, K. et al.: ibid. **57**, 879 (1984)
54. Schneider, W., Zelewsky, A. V.: Helv. Chem. Acta **46**, 1848 (1963)
55. Griffiths, T. R., Scarrow, R. K.: Trans. Farad. Soc. **65**, 1727 (1969)
56. Gruen, G. M., McBeth, R. L.: J. Phys. Chem. **63**, 393 (1959)
57. Griffiths, T. R., Scarrow, R. K.: Trans. Farad. Soc. **65**, 2567 (1969)

58. Griffiths, T. R., Scarrow, R. K.: ibid. **65**, 1427 (1969)
59. Griffiths, T. R., Scarrow, R. K.: ibid. **65**, 3179 (1969)
60. Angell, C. A., Gruen, D. M.: J. Am. Chem. Soc. **88**, 5192 (1966)

CHAPTER C

Thermochromism of Nickel(II) Chelates in Solution

C.I Introduction

The results and discussions in the last chapter may lead to an impression that, although the observed phenomena themselves are quite interesting, most of the theoretical interpretations are far from perfect, and often haunted by numerous ambiguities.

In this and the following chapters we shall discuss a large number of thermochromic and chromotropic solutions, which can be treated in a much more clear-cut way using certain modern concepts in chemistry. The donor-acceptor approach of Gutmann et al. [1, 2] was found to be particularly successfull; chromotropic data can even be used to support such an approach.

The secret of these solutions is very simple; they contain metallic chelates of suitable design, i.e. structural and spectral characteristics which can be easily related to each other, and which are sensitive toward certain types of external influences. We shall explain these situations with some examples of nickel(II) chelates.

C.II Tetrammine-type Chelates: Octahedral-Square Planar Equilibria

In the last chapter it was pointed out that Ni^{2+} forms tetrahedral complexes under certain circumstances. Usually, however, it forms either high-spin octahedral complexes or low-spin square planar complexes. The former are mostly blue, green, or violet, while the latter are yellow, orange or red. Although other geometries and exceptional colors are sometimes observed, they are not numerous. This simple correlation between color and structure is very useful in the study of Ni(II) complexes.

Figure C.1 shows the spectral features of these two types of complexes, with aqueous solutions of $[Ni(H_2O)_6]^{2+}$, $[Ni(en)_3]^{2+}$ and $[Ni(2,3\text{-dimethyl-}2,3\text{-diaminobutane})_2]^{2+}$ (abbreviated as $[Ni(Me_4en)_2]^{2+}$) serving as examples [3, 4]. The first two complexes are octahedral and paramagnetic, and green and violet, respectively. Their spectra consist of three weak bands ($\varepsilon < 10$), resulting from the excitation of an electron from a t_{2g} to an e_g level. Interelectronic repulsions among the eight 3d-electrons occupying these orbitals brings about three kinds of transitions with widely separated energies. The one with the lowest $\tilde{\nu}$ appears in the near infrared, while that with the highest $\tilde{\nu}$ is usually observed in the near ultraviolet. The spectra of the complexes $[Ni(H_2O)_4en]^{2+}$ and $[Ni(H_2O)_2en_2]^{2+}$ lie between those of $[Ni(H_2O)_6]^{2+}$ and $[Ni\ en_3]^{2+}$; all three maxima are shifted regularly to higher $\tilde{\nu}$ with an increase in en.

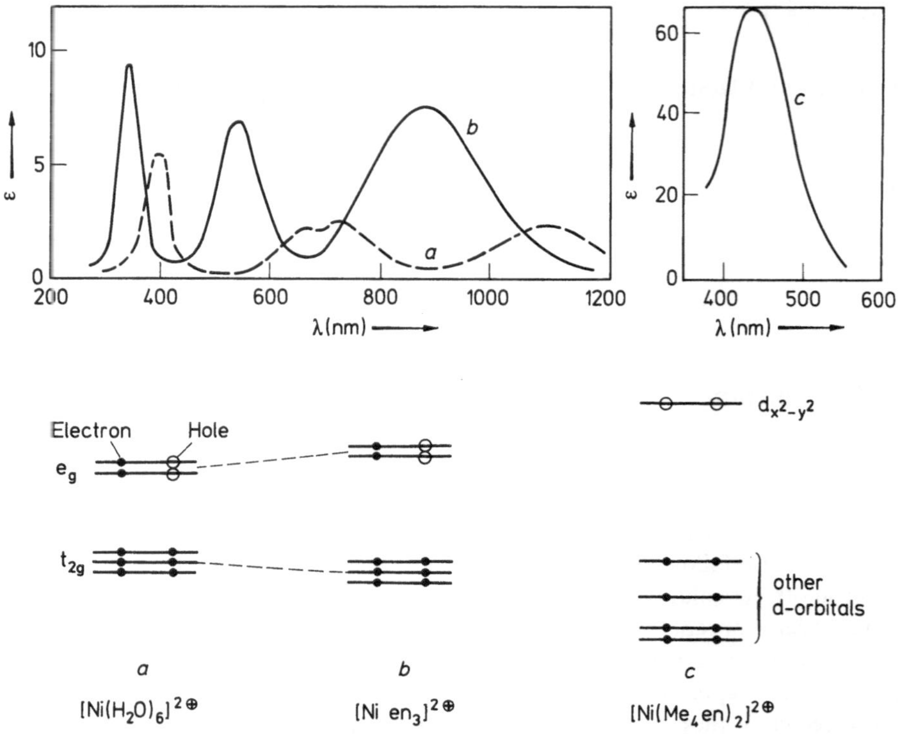

Fig. C-1. Apsorption spectra of typically octahedral complexes ($[Ni(H_2O)_6]^{2+}$ and $[Ni(en)_3]^{2+}$) and a square planar complex ($[Ni(Me_4en)_2]^{2+}$) of $Ni^{2+}(3d^8)$ in aqueous solutions, and sketches of electron distributions in their ground states. Energy scale is only qualitative. (After Cotton and Wilkinson [3], and Basolo et al. [4])

Octahedral Ni(II) complexes of other ligands also show similar spectra. However, the band appearing in the near ultraviolet is often covered by the ligand absorption band, or the CT band, which appears in the same region with a much higher intensity.

On the other hand, when all of the H atoms of the ethylene chain of an en are substituted by methyl groups, the resulting ligand Me_4en becomes very bulky and much more hydrophobic than the unsubstituted en. Thus it cannot form a chelate of the type $[Ni(diamine)_3]^{2+}$ in aqueous solution; a square planar, yellow and diamagnetic complex of the type, $[Ni(diamine)_2]^{2+}$ is readily formed instead. The latter was shown by Basolo and coworkers [4] to lack coordinated water. Its spectrum is composed of a single band at ca. 430 nm with a relatively high intensity ($\varepsilon = 65$). Square planar Ni(II) complexes with other ligands also exhibit such a band at 400–600 nm with $\varepsilon \approx 10^2$. In a square planar complex, the $d_{x^2-y^2}$ orbital confronting the ligands is strongly destabilized by their repulsion, so that the eight 3d-electrons are driven into the remaining four lower orbitals. The excitation of one of these electrons to the vacant $d_{x^2-y^2}$ orbital gives rise to the observed band.

It is interesting to note that the conversion between these two types of complexes can often be observed in terms of chromotropic phenomena, both in the solid state to be described later (Chap. F) and in various kinds of solutions. In 1957 Jørgensen [5] reported that a drastic color change to orange is observed when a large amount of

NaClO$_4$ is added to an aqueous blue violet solution of [Ni trien(H$_2$O)$_2$]$^{2-}$ (trien = triethylenetetramine, NH$_2$(CH$_2$)$_2$NH(CH$_2$)$_2$NH(CH$_2$)$_2$NH$_2$) (Fig. C.2(a)). This is evidently due to the inert salt effect (cf. Chap. B), by which the coordinated water molecules are squeezed out of the coordination sphere:

$$\left[\text{Ni(trien)(H}_2\text{O)}_2\right]^{2\oplus} \xrightarrow{\text{NaClO}_4} \left[\text{Ni(trien)}\right]^{2\oplus} + 2\,\text{H}_2\text{O}$$

blue   orange

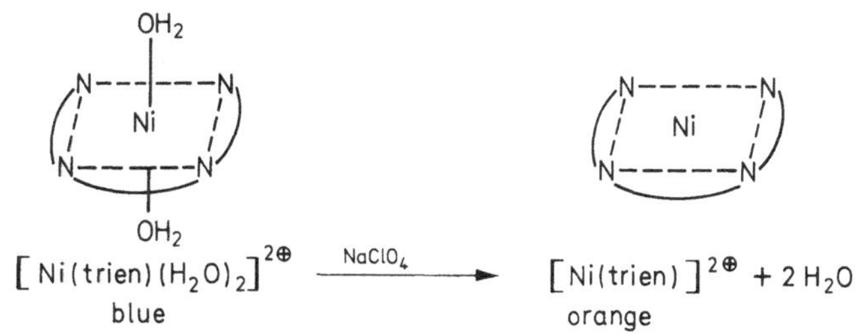

(Eq. C.1)

The structure of [Ni(trien)(H$_2$O)$_2$]$^{2+}$ in solution may be trans, as shown above, or cis. In fact, Jørgensen has suggested the latter structure from spectral evidences. However, his approach may not be conclusive enough, since some dubious data exist. We shall assume the trans structure in this case and in similar cases to be discussed below, because it usually leads to a simple and consistent explanation.

Being interested in Jørgensen's work, Sone and Kato [6] tried to observe the same inert salt effect, and the effect of anhydrous alcohol as the solvent on this type of equilibrium, using en and pn (pn = propylenediamine, NH$_2$CH$_2$CH(CH$_3$)NH$_2$ as diamine ligands. They observed that the blue-violet crystals of the chelates [Ni(diam)$_2$(H$_2$O)$_2$](ClO$_4$)$_2$(diam = en or pn) lose their coordinated water very easily by heating slightly or even during prolonged storage in a desiccator. The orange, anhydrous chelates, which certainly contain the planar [Ni(diamine)$_2$]$^{2+}$ ions, readily absorb water from humid air to become blue-violet again. These anhydrous chelates form blue-violet solutions in water and in anhydrous ethanol. The ethanolic solutions are thermochromic and become reversibly yellowish upon heating, due to a shift in the following equilibrium:

$$[\text{Ni(diam)}_2(\text{EtOH})_2]^{2+} \underset{\text{Cool}}{\overset{\text{Heat}}{\rightleftharpoons}} [\text{Ni(diam)}_2]^{2+} + 2\text{EtOH} \qquad \text{(Eq. C.2)}$$

blue-violet   orange

On the other hand, the aqueous solutions are not thermochromic, indicating that the coordination of water in [Ni(diam)$_2$(H$_2$O)$_2$]$^{2+}$ is stronger than that of EtOH in [Ni(diam)$_2$(EtOH)$_2$]$^{2+}$. However, the addition of a large amount of NaClO$_4$ makes the aqueous solution notably thermochromic, due to the inert salt effect:

$$[\text{Ni(diam)}_2(\text{H}_2\text{O})_2]^{2+} \underset{\text{Cool}}{\overset{+\,\text{NaClO}_4,\text{Heat}}{\rightleftharpoons}} [\text{Ni(diam)}_2]^{2+} + 2\text{H}_2\text{O} \qquad \text{(Eq. C.3)}$$

Some examples of the observed spectral changes are given in Fig. C.2b and c. They show that pn is more effective than en in shifting these equilibria to the right-hand side, both in water and ethanol. As shown above, in the case of the trien chelate, the inert salt effect can change the color of a solution even at room temperature. However, the cooperation of the inert salt and heat is necessary to bring about the same change with en or pn. From all these facts, Sone and Kato deduced that the ease of an octahedral →

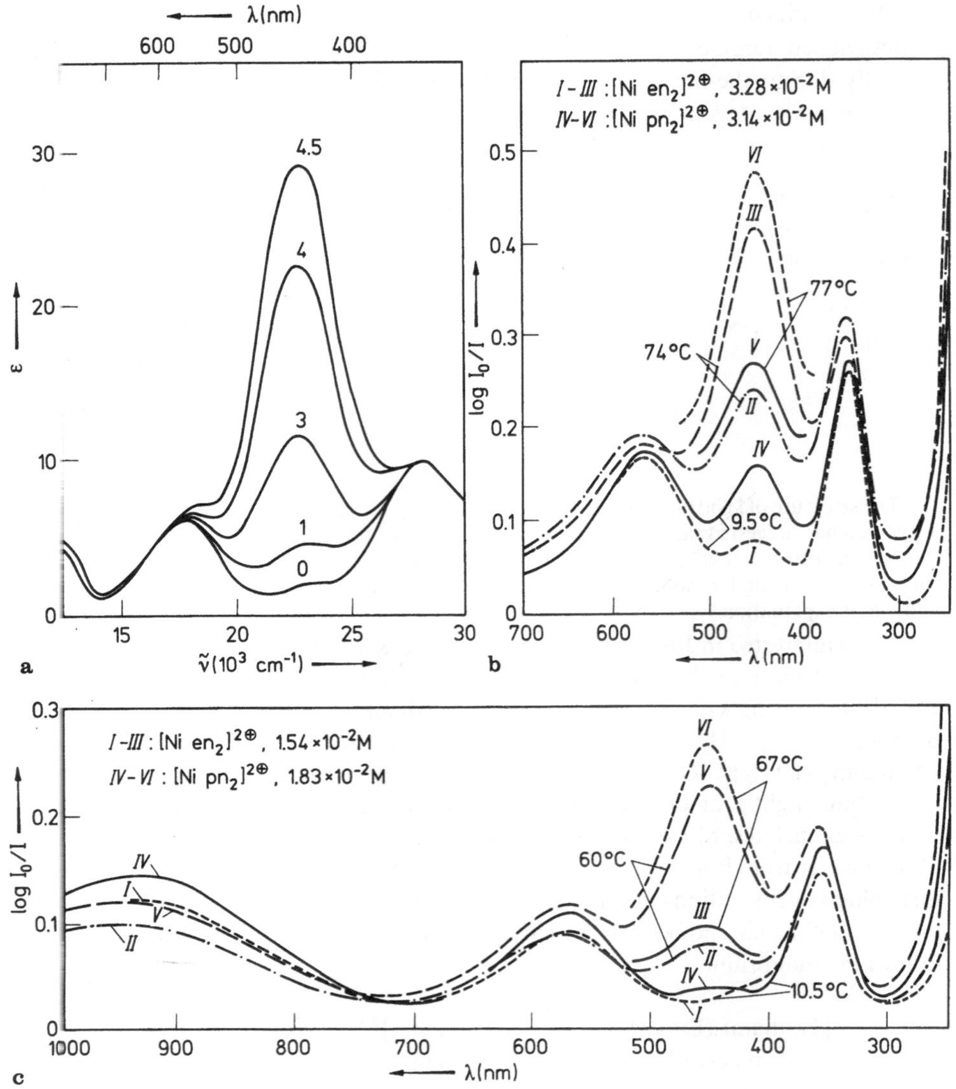

Fig. C-2 a–c. (a): Absorption spectra of $[Ni(trien)(H_2O)_2](NO_3)_2$ (10^{-1} M) in water and aqueous $NaClO_4$ solutions at 19 °C. The numbers on the curves indicate the concentration of $NaClO_4$ (mol dm^{-3}) in each solution (after Jørgensen [5]). **(b)** and **(c):** Absorption spectra of $[Ni(en)_2](ClO_4)_2$ and $[Ni(pn)_2](ClO_4)_2$ in 4 M-$NaClO_4$ solution **(b)** and in ethanol **(c)** at various temperatures. Cell thickness: 10 mm. (After Sone and Kato [6])

square planar (or blue → yellow) change for these NiN_4-type chelates increases in the order of their ligands:

$$\text{en} < \text{pn} < \text{trien} < \text{Me}_4\text{en}$$

i.e. with increasing bulkiness of the alkyl and ethylene substituents in the ligands, which hinder the coordination of solvent molecules.

It should be noted that the steric effect is by no means the only driving force behind these changes in color. The ligand field of en is about 40% stronger per $-NH_2$ (as estimated from the spectra (cf. Fig. C.1)) than that of H_2O. As a result the water molecules in $[Ni(en)_2(H_2O)_2]^{2+}$ are playing a "tough game", in holding out against the strong tendency of the en molecules to displace them out of the coordination sphere and form a low-spin square planar chelate with a higher ligand field stabilization. This marginal stability of the coordinated water molecules is further decreased by the steric effect, bringing about the order mentioned above. In general, it can thus be said that the octahedral → square planar change is favored by (i) the increasing ligand field strength (l. f. s.) of the equatorial ligands, e. g., en or trien, I_{eq}; (ii) the increasing difference between I_{eq} and I_{ax} (the l. f. s. of the axial ligands, e. g., H_2O; $I_{ax} < I_{eq}$), and (iii) the steric repulsion between the axial ligands and bulky groups on the equatorial ligands which makes I_{ax} still weaker.

Similar spectral changes for other $[Ni(diam)_2]^{2+}$ chelates have been reported by a number of investigators. For example, Leussing et al. [7] studied the effects of various alkali metal salts on the equilibrium between $[Ni(meso\text{-}bn)_2(H_2O)_2]^{2+}$ and $[Ni(meso\text{-}bn)_2]^{2+}$ (bn $= NH_2CH(CH_3)CH(CH_3)NH_2$). Farago et al. [8] studied similar equilibria between the chelates of ibn($= NH_2CH_2C(CH_3)_2NH_2$), in water, methanol, and ethanol, and in the presence of various halide ions.

Apparently anomalous results are sometimes observed with certain solvents if the anions possess noticeable coordination ability towards $[Ni(diam)_2]^{2+}$. For example, Nyburg et al. [9] found that the complex $[Ni(meso\text{-}stien)_2(dca)_2]$ (stien $= NH_2CH(C_6H_5)CH(C_6H_5)NH_2$, dca $= CHCl_2COO^-$) is blue and octahedral in $CHCl_3$ or acetone, but becomes remarkably yellow and square planar when water is added to an acetone solution. If, following the idea of Jørgensen [10], we assume that the coordination abilities of the potential ligands in this system increase in the order of acetone < dca^- < H_2O, the latter observation can be explained by considering the following equilibrium:

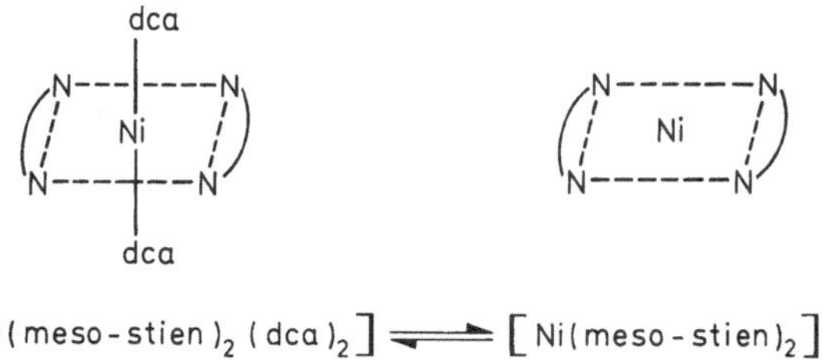

$$[Ni(meso\text{-}stien)_2(dca)_2] \rightleftharpoons [Ni(meso\text{-}stien)_2]^{2\oplus} + 2\,dca^{\ominus}$$

blue yellow

(Eq. C.4)

In acetone, the blue species $[Ni(meso\text{-}stien)_2(dca)_2]$ remains intact. With the addition of water it is ionized (i. e., the dca^- ions are strongly hydrated and drawn out of the coordination sphere) and converted into the yellow complex $[Ni(meso\text{-}stien)_2]^{2+}$, which cannot be effectively hydrated owing to the bulky and hydrophobic phenyl groups above and below the chelate plane. It does not become hydrated even in aqueous solutions.

One can also say that, since the axial coordination sites are effectively shielded by such phenyl groups, the equilibrium (Eq. C.4) is solely governed by the ability of the solvent to solvate the anion (dca^-) and pull it away from the Ni^{2+} ion, i. e., by the acceptor number (AN; cf. Chap. D) of the solvent. The AN of water is very high (54.8), but that of acetone (12.5) is very low, so that the equilibrium (Eq. C.4) is virtually shifted from one side to the other.

Nyburg et al. analyzed their results with such a view in mind. For example, they found that ca. 1/3 of the complex exists as $[Ni(meso\text{-}stien)_2]^{2+}$ in ethanol at room temperature, while the rest is in the form of trans-$[Ni(meso\text{-}stien)_2(dca)_2]$ [11]. It is also interesting to note that they obtained blue and yellow crystals from aqueous-ethanolic solutions of different concentrations. The blue crystals contain only the octahedral species, while the yellow crystals contain 1/3 and 2/3 of the blue and yellow species, respectively [12]. They also observed that the yellow crystals are much more difficult to prepare, when the racemic stien is used instead of the meso-form. The yellow crystals obtained form blue aqueous solutions, due to the arrangement of the phenyl groups which do not severely hinder the axial solvation (cf. Fig. C.3).

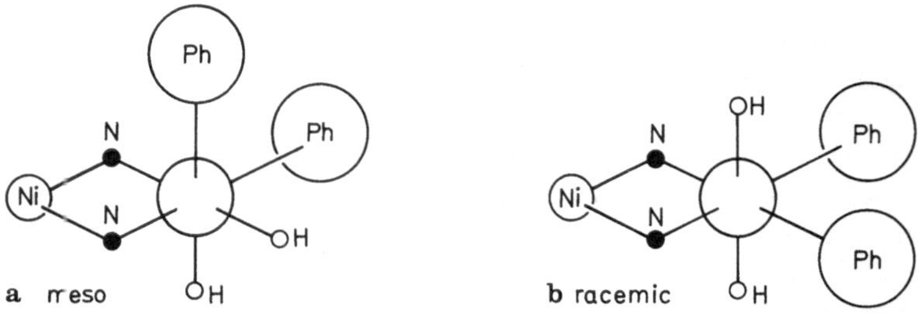

Fig. C-3 a, b. Steric effects in the $[Ni(diamine)_2]^{2+}$-type chelates of meso- and racemic-stien. Viewed along the C–C axis, their chelate rings will appear as in **a** or **b**. In **a**, one of the bulky phenyl groups occupies an axial site, hindering the coordination of other ligands; since there is another chelate ring with an axial phenyl group, Ni^{2+} is shielded from above and below. In **b**, on the other hand, both of the phenyl groups can occupy the equatorial sites, so that axial coordination to Ni^{2+} is much easier

Gillard and Sutton [13–15] studied the blue-to-yellow transition for Ni(meso-stien)$_2(CH_3COO)_2$ in numerous mixtures of water with organic solvents, involving various alcohols, amides, glycol ethers and poly(ethyleneoxide). Sometimes urea and different proteins were also added. The formation of the yellow species is enhanced by an increase in the amount of water in solution. They tried to correlate the ratio R = (yellow form)/(blue form) with the composition and the structure of the solvents. Their approach is unique and interesting. For example, they showed that the addition of urea, a well-known structure-breaking agent for water, increases the activity of water in water-methanol mixtures. The resulting solutions of the nickel chelate are more yellowish. On the other hand, the addition of urea to aqueous solutions of $[Ni(trien)(H_2O)_2]^{2+}$ and $NaClO_4$ causes these solutions to become more bluish by the same effect. Ionic hydration is promoted by urea in both cases, but in different ways, shifting the equilibria in opposite directions (compare Eqs. C.1 and C.4 to see how this can occur).

In the latter case, it is also conceivable that the highly polar molecules of urea can solvate the ions of $NaClO_4$ to some extent, displacing some of the water in their coordination spheres and relieving the chelate from dehydration.

Furthermore, they found that many water-organic solvent mixtures exhibit a characteristic inflection point on the curve relating log R and log S (S = total concentration (mol dm^{-3}) of water and organic solvent) at the composition where the ultrasonic absorption or viscosity of the mixture shows a maximum. This change in color can thus serve as a tool of diagnosis in the elucidation of solvent structures.

The complicated color changes observed with the nickel(II) chelates of stien and phenen($= NH_2CH(C_6H_5)CH_2NH_2$) (often called Lifschitz's chelates after their discoverer [16–18]) and of many other related derivatives of en will also be referred to in connection with the thermochromism of solids (Chap. F). It may be added that Curtis and House [19] have shown that even $[Ni(trien)(py)_2]^{2+}$ exhibits thermochromic equilibria, losing pyridine of high coordination ability in water, methanol, and other polar solvents, changing color from violet to yellow. They also found that the chelate $[Ni(trien)(py)(H_2O)]Cl_2$ changes its color from blue to bluish green upon heating in methanol. In this case the Cl$^-$ ions seem to be coordinated to the yellow species $[Ni(trien)]^{2+}$ produced in hot solution.

For NiN$_4$-type chelates with a single polydentate ligand, such as $[Ni(trien)]^{2+}$, the work of Iwasaki et al. [20] may be of interest. They used the ligand "bapp" with the following structure,

$$NH_2CH_2CH_2CH_2N\underset{\diagdown CH_2CH_2\diagup}{\overset{\diagup CH_2CH_2\diagdown}{}}NCH_2CH_2CH_2NH_2$$

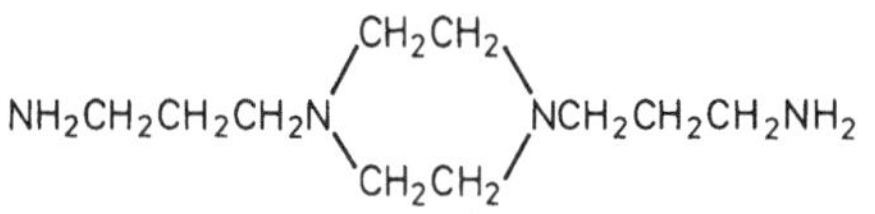

and discovered that, in certain organic solvents (acetonitrile and DMF), its orange nickel(II) complex of the structure,

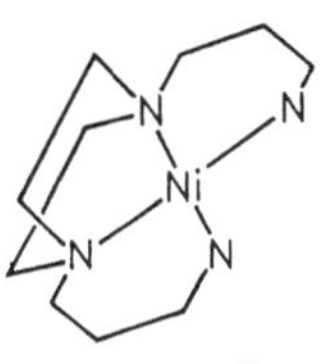

$[Ni(bapp)]^{2+}$

becomes bluish upon cooling down to ca. $-50°C$. The spectral changes indicate the formation of octahedral species at lower temperature. They also noted that, when Cl$^-$ or Br$^-$ ions are present in such solutions, another thermochromic equilibrium of the type,

$$[Ni(bapp)]^{2+} + Cl^- \underset{Cool}{\overset{Heat}{\rightleftharpoons}} [Ni(bapp)Cl]^+ \qquad\qquad (Eq.\,C.5)$$

between the halide-containing 5-coordinate species discovered by Gibson and McKenzie [21] and the original square planar species is established. As a result the spectral changes observed upon heating or cooling become even more complicated.

More recently, Mochizuki et al. [22] studied in detail the thermochromism of four macrocyclic chelates *1* to *4* containing a pyridine ring in aqueous solutions.

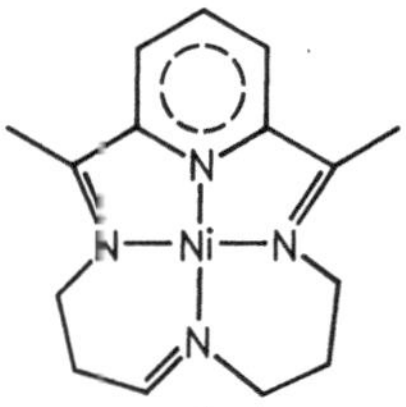

1

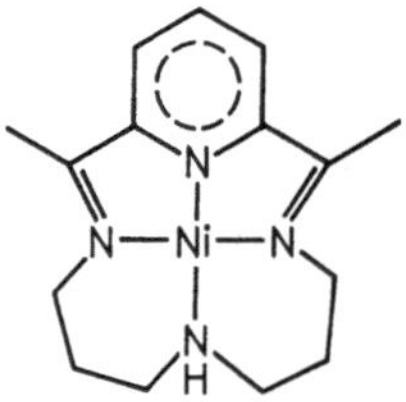

2

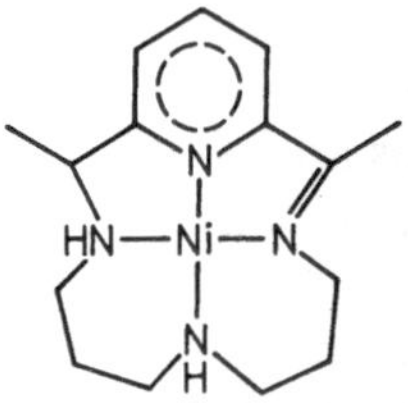

3

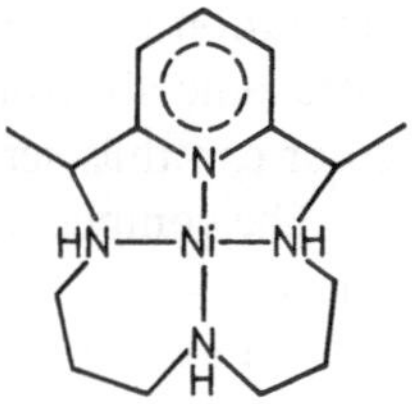

4

Spectral changes similar to those given in Fig. C.2 are observed. The value of K, ΔH and ΔS for the equilibrium,

$$[NiL(H_2O)_2]^{2+} \rightleftharpoons [NiL]^{2+} + 2H_2O,$$

are given in Table C.1.

Table C.1. K, ΔH and ΔS values of the thermochromic equilibrium for the chelates *1–4* (After Mochizuki et al. [22])

Chelate[a]	K[b,c]	ΔH/kJ(mol^{-1})	ΔS/J(K^{-1} mol^{-1})
1	1.92	23.2	82.0
2	0.28	21.2	59.4
3	2.04	21.6	77.0
4-a(meso form)[d]	3.57	20.2	77.0
4-b(rac. form)[d]	2.44	20.4	74.9

[a] Perchlorates. [b] $I = 0.1(NaClO_4)$, 303–304K.
[c] $K = [NiL]^{2+}/[NiL(H_2O)_2]^{2+}$. [d] C-meso and C-racemic, respectively.

In all of these chelates, the equatorial ligand field strength (I_{eq}; see above) is very strong, so that even at room temperature a high percentage is in the form of the yellow complex $[NiL]^{2+}$. One can anticipate that I_{eq} increases from **4** to **1**, i.e., with the development of the π-system in the ligand, since the possibility of π-type coordination (back donation) also increases in the same direction. On the other hand, the steric hindrance towards axial ligands increases in the opposite order, since the periphery of the macrocyclic ring becomes less planar in going from **1** to **4**. The unusual stability order of $[NiL(H_2O)_2]^{2+}$ for these chelates is a compromise of these two tendencies. A strong inert salt effect was also observed; by adding $NaClO_4$ (6 M), the blue species practically disappears. Observations with macrocyclic NiN_4-type chelates were also reported by other investigators [23–24].

C.III β-Diketonate Chelates: Monomer-Polymer Equilibria

The (octahedral → square planar) type of thermochromism encountered with nickel(II) chelates is also observed with chelating ligands that are bound to the metal by oxygen atoms. A simple but interesting example is shown by $Ni(dike)_2$($dike^- =$ β-diketonate ion, e.g., acetylacetonate ion ($acac^-$)), which has been studied extensively by Cotton and Fackler [25–26].

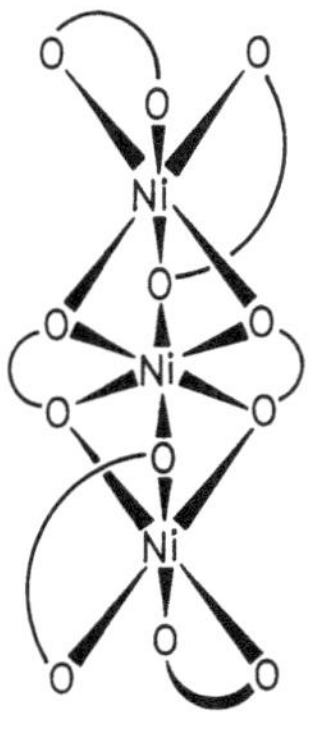

Fig. C-4. Structure of the trimer, $[Ni_3(acac)_6]$, found in solid $Ni(acac)_2$. (After Bullen et al. [27])

The structure of the green solid, $Ni(acac)_2$, was shown by X-ray analysis to be a trimer as illustrated in Fig. C.4 [27]. All of the Ni^{2+} ions are octahedrally surrounded by the ligands. In polar solvents such as alcohols, this trimer dissociates into octahedral monomers containing two solvent molecules.

$$[Ni(acac)_2]_3 + 6Solvent \longrightarrow 3[Ni(acac)_2(Solvent)_2] \qquad \text{(Eq. C.6)}$$

In relatively non-polar solvents they dissolve only partially, seemingly without dissociation. However, when the methyl groups (or the central H atom) in the ligand are substituted by more bulky groups such as tert-butyl or phenyl, the formation of the trimer is sterically hindered, and a red, square planar monomer is formed instead. Thus, $Ni(dipm)_2$ (dipm = dipivaloylmethanate, with two tert-butyl groups instead of methyl in acac) is monomeric and red, both in the solid state and in non-polar solutions. Chelates such as $Ni(dibm)_2$ (dibm = diisobutyrylmethanate, with two isopropyl) or $Ni(ppd)_2$(ppd = 3-phenyl-2,4-pentanedionate, with a C_6H_5 group instead of the central H atom) form either red or blue crystals, or sometimes both, depending on the method of preparation. Both the monomeric and trimeric forms coexist in solution. The equilibrium in such solutions,

$$\underset{\substack{\text{trimer} \\ \text{blue or green}}}{[Ni(dike)_2]_3} \rightleftharpoons \underset{\substack{\text{monomer} \\ \text{red}}}{3[Ni(dike)_2]} \qquad \text{(Eq. C.7)}$$

is shifted to the right-hand side by heating, so that a characteristic thermochromism is observed. The spectral changes observed in a toluene solution of $Ni(dibm)_2$ are shown in Fig. C.5.

It can be seen that the species with a band at ca. 550 nm, which appears upon heating, corresponds to the monomer. By analyzing the curves, it was found that the value of K for the dissociation in this solution increases from $6.67 \cdot 10^{-5}$ at 25 °C to $4.48 \cdot 10^{-4}$ at 50 °C, with $\Delta H = 62.8\ kJmol^{-1}$ and $\Delta S = 130\ JK^{-1}\ mol^{-1}$. It is further interesting to note that even $Ni(acac)_2$ itself is thermochromic; when heated in diphenylmethane up to ca. 200 °C, it becomes reversibly red, showing spectral changes given in Fig. C.6.

Further studies of related chelates, which are also thermochromic, were performed by Yoshida et al. [28]. They indicated that dimeric octahedral species can also form in

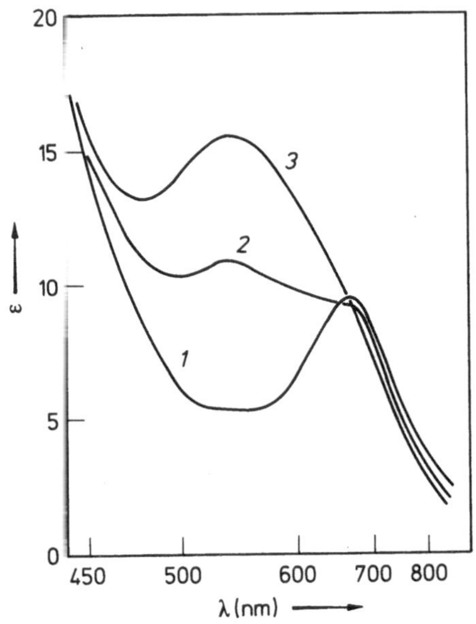

Fig. C-5. Absorption spectra of $Ni(dibm)_2$ in toluene. Conc.: 4.365×10^{-2} M. Temp.: (*1*), 2.0 °C; (*2*), 34.0 °C; (*3*), 52.3 °C. (After Cotton and Fackler [25])

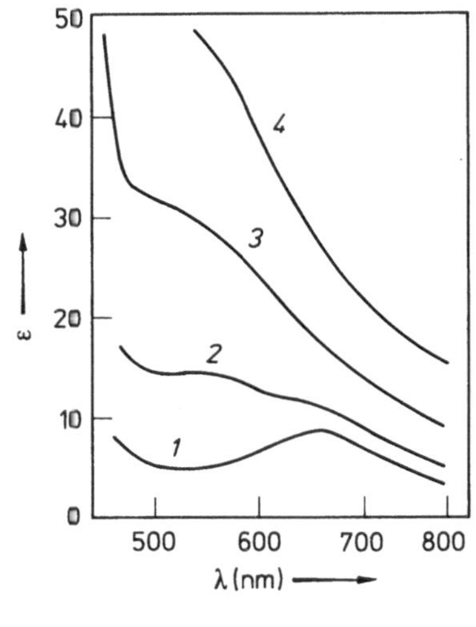

Fig. C-6. Absorption spectra of $Ni(acac)_2$ in $CH_2(C_6H_5)_2$. Conc.: 2.08×10^{-2} M. Temp.: (*1*), 80 °C; (*2*), 120 °C; (*3*), 160 °C; (*4*), 200 °C. (After Fackler and Cotton [26])

solution when the bulkiness of the alkyl groups lies between that of the methyl groups in acac and that of the tert-butyl groups in dipm.

Since the ligand field of O-donor ligands is generally weaker than that of N-containing ones, it is to be expected that a square planar NiO_4-type chelate will be much less stable than a NiN_4-type chelate of similar geometry and will tend to be converted into an octahedral NiO_6-structure by combining with additional ligands, either by solvation or by polymerization. The use of a non-polar solvent eliminates the former possibility, while the use of bulky substituents suppresses the latter. The observed thermochromism is the result.

It may be noted that square planar NiO_4 chelates, e.g., $Ni(dipm)_2$, are usually red (λ_{max}: 500–550 nm). The analogous Ni-N_4 chelates are usually yellow to orange (λ_{max}: 400–450 nm), corresponding to an increase of the ligand field strength. It can be seen that the λ_{max} values of the square planar NiN_2O_2 chelates discussed in the next section generally lie between these two limits.

The recent results of Chamberlain and Drago [29] are of interest in connection with these topics. They prepared the chelates $Ni(hpd)_2$, $Ni(hed)_2$ and $Ni(thdd)_2$, in which the methyl groups of acac have been substituted by long chain alkyls, i.e., n-heptyl, n-nonyl and tert-nonyl, respectively. Some of their physical properties are listed in Table C.2.

Table C.2. Physical properties of the long-chain derivatives of $Ni(acac)_2$ (After Chamberlain and Drago [29])

Chelate	Stable form at room temp.	μ_{eff}/B.M.	t/°C[a]
$Ni(acac)_2$	Green trimer, solid	3.3	–
$Ni(hpd)_2$	Green trimer, liquid[b]	3.3	17
$Ni(hed)_2$	Purple monomer, solid	0	42
$Ni(thdd)_2$	Purple monomer, liquid[c]	0	–[d]

[a] Monomer-trimer transition temperature. [b] When kept below 0 °C, this chelate is slowly converted into its solid purple monomer. Upon heating it, the formation of the solid green trimer begins at 17 °C, which melts into a viscous liquid at 19 °C. [c] See text. [d] When kept at ca. −16 °C it solidifies; the m.p. of the solid is, however, much higher (2 °C).

It is notable that the square planar monomers of these chelates are colored purple, probably due to a small red shift of their absorption bands, while the trimers are green as in the case of $Ni(acac)_2$. The chelate $Ni(thdd)_2$ only forms the monomer, evidently due to the steric effect of the tert-alkyl groups. $Ni(hpd)_2$ and $Ni(thdd)_2$ are liquids, while the melting point of $Ni(hed)_2$ is also very low. The thermal behavior of $Ni(hed)_2$ is especially peculiar; upon heating it first polymerizes and then depolymerizes, as shown schematically below:

$$\underset{\text{(monomer)}}{\text{purple solid}} \xrightarrow{>42\,^\circ C} \underset{\text{(trimer)}}{\text{green solid}} \xrightarrow{<53\,^\circ C} \underset{\text{(trimer)}}{\text{green liquid}} \xrightarrow{>53\,^\circ C} \underset{\text{(monomer)}}{\text{purple liquid}}$$

(green liquid (trimer) → purple solid (monomer): Room temp. (25 °C, 30 min.))

The thermal behaviour of $Ni(hpd)_2$ is also peculiar (cf. footnote b in Table C.2).

C.IV Mixed Diamine-β-Diketonate Chelates: Solvatochromic and Thermochromic Equilibria Governed by the Coordinating Power (Donor Number) of the Solvent

C.IV.1 Introduction

The facts described in Sections C.II and C.III lead to the idea that, if it is possible to hybridize $[Ni(diam)_2]^{2+}$ and $Ni(dike)_2$, i.e., to prepare the mixed chelates $[Ni(diam)(dike)]^+$, they should exhibit highly interesting chromotropic behavior. In fact, Fukuda and Sone prepared this type of mixd chelate for the first time, using tmen

($(CH_3)_2NCH_2CH_2N(CH_3)_2$) as diam, and acac and bzac(=benzoylacetonate ion; see below) as dike, and confirmed this expectation in 1972 [30]. Since then, a large number of similar chelates have been prepared and comparative studies of their solvatochromic and thermochromic properties have been carried out [31–33]. This topic will be presented in this section.

The abbreviations shown in Table C.3 will be used hereafter to designate various ligands used in these studies.

Table C.3. Abbreviations used to designate various diamine(diam) and β-diketonate(dike) ligands (cf. Fig. C.7, formula I)

A: Diamine ligands with different substituents R_1–R_4 on N atoms:					
Name	Abbreviation	R_1	R_2	R_3	R_4
N,N N′,N′-tetramethyl-en[a]	tmen	Me	Me	Me	Me
N,N N′,N′-tetraethyl-en	teen	Et	Et	Et	Et
N,N′-dimethyl-en	sym-dmen	Me	H	Me	H
N,N-dimethyl-en	unsym-dmen	Me	Me	H	H
N,N′-diethyl-en	sym-deen	Et	H	Et	H
N,N-diethyl-en	unsym-deen	Et	Et	H	H
B: β-Diketonate ligands with different substituents R_5 and R_6 on CO groups:					
Name	Abbreviation	R_5		R_6	
acetylacetonate	acac	Me		Me	
benzoylacetonate	bzac	Me		Ph	
dibenzoylmethanate	dibm	Ph		Ph	
trifluoroacetylacetonate	tfac	Me		CF_3	
hexafluoroacetylacetonate	hfac	CF_3		CF_3	
dipivaloylmethanate	dipm	tert-Bu		tert-Bu	

[a] en stands for ethylenediamine.

C.IV.2 Chelate Perchlorates and Tetraphenylborates

The structures of the mixed complexes which were extensively studied are shown in Fig. C.7.

The type-I chelates, i.e., the perchlorates and tetraphenylborates of the chelate cation $[Ni(diam)(dike)]^+$, are all red, planar and diamagnetic, while those of the types

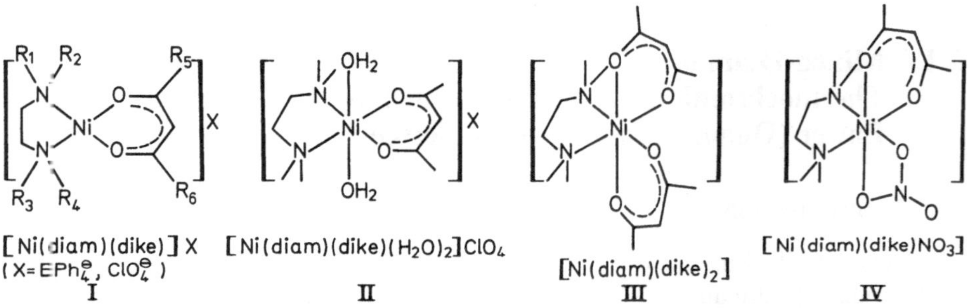

Fig. C-7. Representative types of mixed diamine-β-diketonato chelates of nickel(II) studied by Fukuda, Sone and collaborators [30–33]

II to IV are all blue or green, octahedral and paramagnetic. Most of these complexes are quite easy to prepare and, rather amazingly, are soluble in a very large number of organic solvents, ranging from highly non-polar 1,2-dichloroethane(DCE) to alcohols, DMF, DMSO and pyridine.

The color of the solutions of the chelates of type I strongly depends on the nature of the solvent; in a typically non-coordinating solvent such as DCE, the solution is red, while in a highly coordinating solvent such as DMF, DMSO or pyridine, blue or green solution is obtained. In solvents of intermediate coordinating power, such as alcohols or acetone, the solutions exhibit intermediate colors, which strongly depend upon the temperature. At higher temperatures they are red, whereas they are blue or green at lower temperatures. Thus, it is clear that there exists an equilibrium of the type:

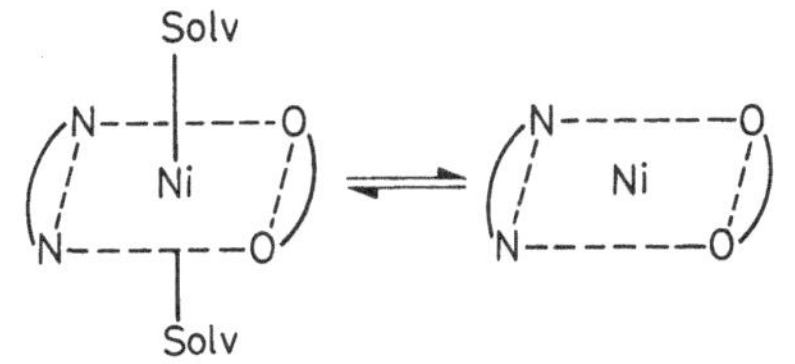

$$[\mathrm{Ni(diam)(dike)(Solv)_2}]^{\oplus} \underset{\text{Cool}}{\overset{\text{Heat}}{\rightleftharpoons}} [\mathrm{Ni(diam)(dike)}]^{\oplus} + 2\,\mathrm{Solv}$$

blue or green — red

(Eq. C.8)

If the coordination strength of the solvent is low, this equilibrium is shifted to the right hand side to yield a red solution. If it is high, a solvated octahedral species predominates and the solution is blue or green. A thermochromic mixture of the blue and red species is obtained with solvents of intermediate coordinating strength. As in the case of $[Ni(diam)_2]^{2+}$- and $Ni(dike)_2$-type chelates, the square planar (i.e., red) species tends to predominate at higher temperatures.

The dihydrates of the perchlorates (type II) are sometimes formed when the red anhydrous perchlorates are exposed to humid air. However, the $Ni\text{-}(OH_2)$ bonds in them are seemingly weak, so that most of the coordinated water is driven out of the coordination sphere when these hydrates are dissolved in organic solvents (even in DCE). The resulting solutions exhibit colors which are similar to those of the solutions of anhydrous perchlorates.

It is now convenient to introduce the concept of donor numbers (DN) proposed by Gutmann et al. in the 1960's [1, 2]. This concept has been widely used in the study of equilibria and reactions in organic solvents with remarkable success. The donor number is a measure of the donor ability, or Lewis basicity, of a solvent molecule, measured on the basis of the amount of heat evolved when the same solvent is added to a solution of $SbCl_5$ in DCE. $SbCl_5$ is a trigonal bipyramidal molecule. Since a coordinate bond between DCE and $SbCl_5$ can be reasonably ruled out, we can expect the molecular shape of the latter to be retained in DCE. A polar solvent (D), added to such a solution, will be coordinated to the antimony atom through its electron pair, forming a solvated 6-coordinate complex as shown diagramatically below.

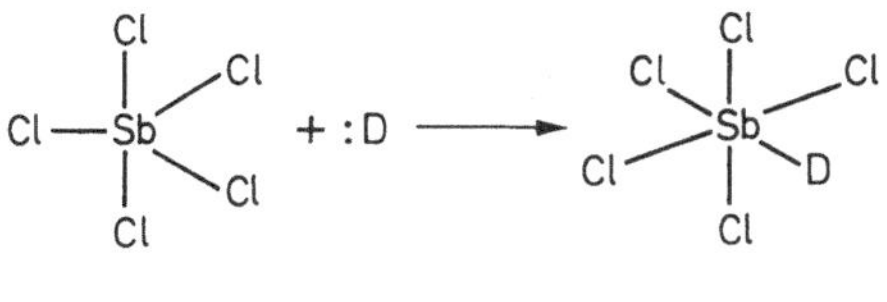

The heat evolved in the course of this reaction ($-\Delta H$), measured in kcal mol^{-1} (1 kcal = 4.18 kJ), is what Gutmann called DN. It can be taken to be a measure of the σ-donor strength of the solvent, i.e., its power to form a σ-type coordinate bond with a positively charged particle. Table C.4 lists the values of DN of various solvents, recently given in Mayer's review [34], together with their acceptor numbers (AN) to be discussed later, in addition to some other polarity parameters.

Table C.4. Acceptor numbers AN, donor numbers DN, dielectric constants D and dipole moments μ (Debye) of various solvents at 25 °C. (After Mayer [34])

Solvent	AN	DN	D	μ
Acetic acid	52.9	–	6.2	1.75
Acetic anhydride	–	10.5	20.7	2.82
Acetone	12.5	17.0	20.7	2.86
Acetonitrile	18.9	14.1	36.0	3.44
Acetyl chloride	–	0.7	15.8	2.71
Benzene	8.2	0.1	2.3	0.00
Benzonitrile	15.5	11.9	25.2	4.05
Benzoyl chloride	–	2.3	23.0	3.26
Benzoyl fluoride	–	2.0	22.7	–
Benzyl cyanide	–	15.1	18.4	3.50
n-Butyl alcohol	36.8	–	17.5	1.75
tert-Butyl alcohol	27.1	–	12.5	1.66
γ-Butyrolactone	17.3	–	39.1	4.03
n-Butyronitrile	–	16.6	20.3	3.57
iso-Butyronitrile	–	15.4	20.4	3.61
Carbon tetrachloride	8.6	0.0	2.2	0.00
Chloroform	25.1	–	4.7	1.15
1,2-Dichloroethane (DCE)	16.7	0.0	10.1	1.75
Dichloroethylene Carbonate	–	3.2	31.6	3.47
Dichloromethane	20.4	–	8.9	1.57
N,N-Diethylacetamide	–	32.2	–	3.75
Diethylamine	9.4	–	3.6	1.11
Diethylene glycol dimethyl ether (Diglyme)	9.9	–	–	–
Diethyl ether	3.9	19.2	4.2	1.25
N,N-Diethylformamide (DEF)	–	30.9	–	–
N,N-Dimethylacetamide (DMA)	13.6	27.8	37.8	3.81
N,N-Dimethylformamide (DMF)	16.0	26.6	36.7	3.86
Dimethylsulfoxide (DMSO)	19.3	29.8	46.7	3.90
N,N-Dimethylthioformamide	18.8	–	47.5	4.37
Dioxane	10.8	–	2.2	0.45
Ethanol	37.9	–	24.3	1.70
2-Aminoethanol	33.7	–	37.7	2.27
Ethyl acetate	9.3	17.1	6.0	1.88
Ethylene carbonate	–	16.4	89.6	4.87
Ethylenediamine (en)	20.9	55.0	14.2	1.90
Ethylene glycol dimethyl ether (Glyme)	10.2	–	7.0	–
Formamide	39.8	24	109.5	3.37
Formic acid	83.6	–	58.5	1.52
Hexamethylphosphotriamide (HMPA)	10.6	38.8	29.6	4.48
n-Hexane	0.0	0.0	1.9	0.00
Methanesulfonic acid	126.3	–	–	–
Methanol	41.5	19.1	32.6	1.70
Methyl acetate	10.7	16.5	6.7	1.69

Solvent	AN	DN	D	μ
N-Methyl-ε-caprolactam	–	27.1	–	–
N-Methylformamide (NMF)	32.1	–	182.4	3.86
N-Methyl-2-pyrrolidone	13.3	27.3	33.0	4.10
N-Methyl-2-thiopyrrolidone	17.7	–	47.5	–
Morpholine	17.5	–	7.3	1.58
Nitrobenzene	14.8	4.4	34.7	4.03
Nitromethane (NM)	20.5	2.7	36.7	3.57
Propanediol-1,2-carbonate	18.3	15.1	65.0	4.98
n-Propyl alcohol	37.3	19.6	20.1	1.67
iso-Propyl alcohol	33.6	–	18.3	1.70
Propionitrile	–	16.1	27.7	4.03
Pyridine (py)	14.2	33.1	12.3	2.37
Tetrachloroethylene carbonate	–	0.8	9.2	–
Tetrahydrofuran (THF)	8.0	20.0	7.4	1.75
Tetramethylenesulfone	19.2	14.8	43.3	4.81
Tetramethylurea (TMU)	–	29.6	23.5	–
Tributyl phosphate	9.9	23.7	6.8	3.07
Triethylamine	1.4	61.0	2.4	0.79
Trifluoroacetic acid	105.3	–	8.3	2.28
Trifluoroethanol	53.3	–	26.7	2.03
Trifluoromethanesulfonic acid	131.7	–	–	–
Trimethylphosphate	16.3	23.0	20.6	3.02
Water	54.8	16.4	78.5	1.84

If we arrange a number of these polar solvents in test tubes in increasing order of their DN values, and add a small amount of a type-I chelate to each test tube, we can observe an interesting spectrum of colors from red to blue or green. It is more interesting to heat all of the test tubes in a large water bath at the same time. The blue and red ones at the ends remain nearly unchanged, but those of intermediate colors becomes more and more reddish with an increase in temperature and a decrease in DN. The opposite changes are observed when the tubes are cooled.

This observation was originally reported by Fukuda and Sone using several common solvents [31]. Recently Soukup made the same observation with [Ni(tmen)(acac)]BPh_4 using 11 solvents [35]. He confirmed the view held by Fukuda and Sone, and stated that this chelate can serve as a convenient indicator of the DN of a solvent. If DN is below 6, the solution is red, while if it is above 25, the solution is greenish blue. The thermochrmomic color change occurs between these extremes.

Detailed spectral observations of these solutions have also been made. Figures C.8 to C.10 show some of the data. They very clearly show the appearance and disappearance of square planar and octahedral species as a function of the DN and the temperature, confirming the existence of the above-mentioned equilibrium (Eq. C.8) from every point of view.

Recentry, Ogino et al. studied these spectral changes more quantitatively and obtained the values of K, ΔH and ΔS listed in Table C.5 [36]. This table contains several points of interest. First of all, the value of K increases with the decrease of DN as expected, but there are two exceptions. Firstly, for most chelates, pyridine (DN = 33) shows a value of K which is of the same order as that of DMSO (DN = 30). If dike = dipm, it is even much higher than the latter. Secondly, CH_3CN (DN = 14) shows a lower value of K than most alcohols and acetone (DN = 17). Both facts are probably

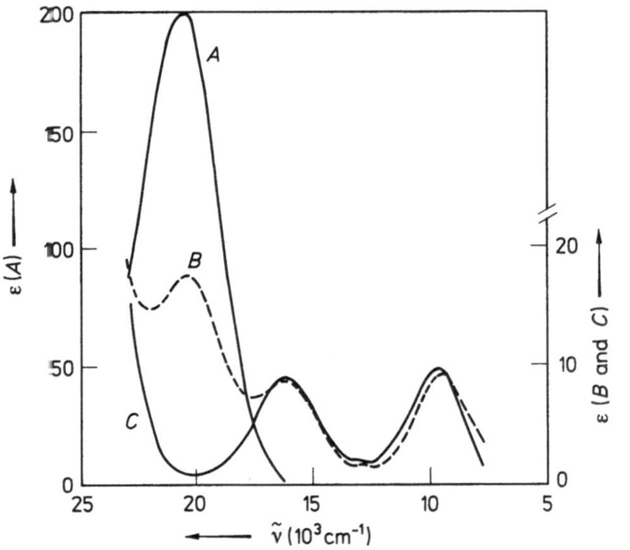

Fig. C-8. Absorption spectra of [Ni(tmen)(bzac)] $B(C_6H_5)_4$ in DCE (*A*), acetone (*B*) and DMF (*C*) at room temperature. (Taken from Fukuda and Sone [30])

related to the steric repulsion that exists between the ligands and solvent molecules. Since a number of bulky alkyl groups on the diamine ligand protrude above and below the chelate plane, it is conceivable that the coordination of the pyridine ring is hindered more strongly than that of DMSO, which is coordinated at its oxygen end. The higher DN of pyridine will be partially cancelled by this effect. In the case of dipm, the bulkiness of the R_5 and R_6 groups on dike (cf. Fig. C.7) makes the situation still worse, favoring DMSO over pyridine. On the other hand, the linear and slim CH_3CN molecule does not suffer such problems. It can slip between the protruding groups quite easily, so that its coordination is favored.

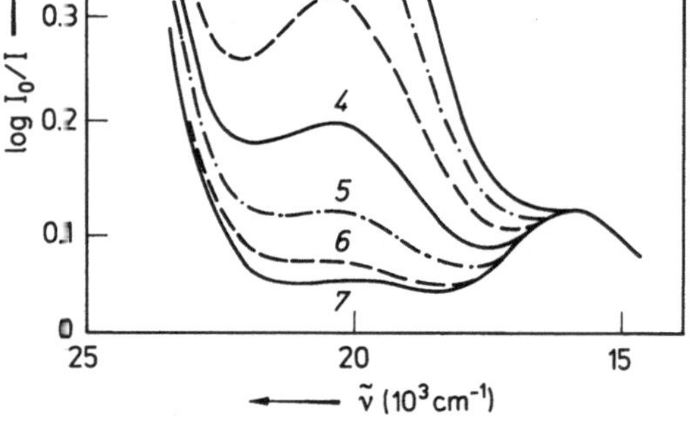

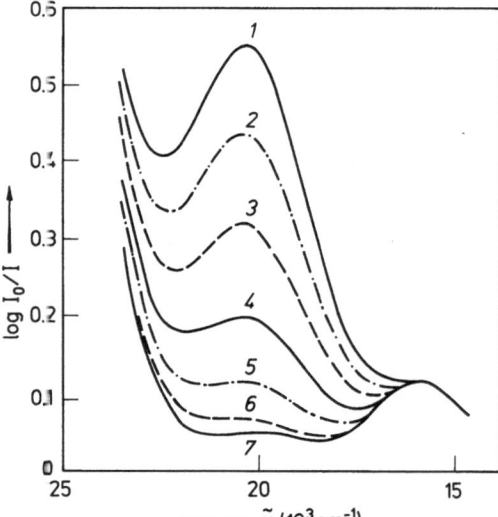

Fig. C-9. Absorption spectra of $[Ni(tmen)(bzac)]ClO_4$ in n-PrOH at various temperatures: (*1*) 73 °C; (*2*) 67 °C; (*3*) 59 °C; (*4*) 47 °C; (*5*) 36 °C; (*6*) 25 °C; (*7*) 11 °C. (Conc.: 1.61×10^{-3} M). (Taken from Fukuda and Sone [30])

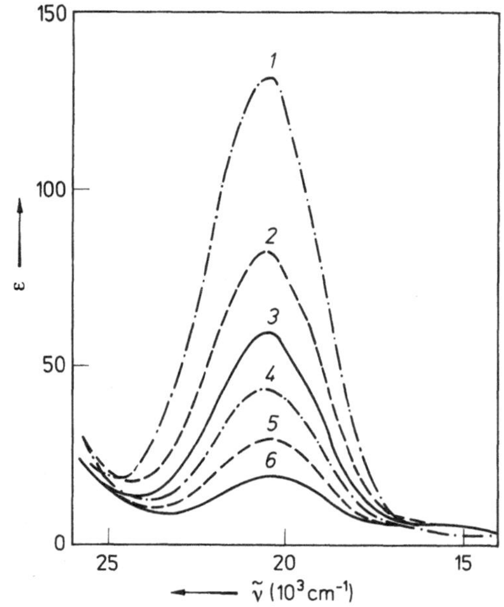

Fig. C-10. Absorption spectra of $[Ni(tmen)(acac)]ClO_4$ in $CHCl_3-$n-PrOH at room temperature (ca. 20 °C). The V/V ratio $[(CHCl_3) / (CHCl_3 + \text{n-PrOH})] \times 100$ in each solvent is (*1*) 100; (*2*) 95; (*3*) 85; (*4*) 60; (*5*) 40; (*6*) 20. (After Fukuda et al. [30])

Table C.5. Equilibrium constants and thermodynamic data of the system $[Ni(tmen)(dike)]^+$ in DCE containing various polar solvents. (After Ogino et al. [36])

A: Equilibrium constants (K) of the reaction $[Ni(tmen)(dike)(Solv)_2]^+ \rightleftharpoons [Ni(tmen)(dike)]^+ + 2Solv$ in DCE at 25 °C:

Solvent	dibm	bzac	acac	dipm
DMSO	7.7×10^{-7}	6.3×10^{-6}	5.3×10^{-5} [a]	3.7×10^{-5}
py	1.0×10^{-6}	2.2×10^{-5}	7.7×10^{-5}	1.2×10^{-3}
CH_3CN	5.9×10^{-2}	1.3×10^{-1}	4.2×10^{-1}	3.8×10^{-1}
MeOH	–	1.1	–	–
EtOH	–	2.8	–	–
n-PrOH	2.3	2.9	17.2	20.8
Acetone	4.6	–	–	33.3

B: Thermodynamic parameters for the above reaction with n-PrOH:

Parameter	dibm	bzac	acac	dipm
$\Delta H/kJ\ mol^{-1}$	59	56	49	35
$\Delta S/JK^{-1}\ mol^{-1}$	204	199	183	145

[a] 35 °C.

In general, it can be seen that the values of K increase in the order dibm → bzac → acac → dipm, in nearly all of the solvents tried. Fukuda and Sone [27] have added two more ligands to this series, making it:

$$\text{hfac} \rightarrow \text{tfac} \rightarrow \text{dibm} \rightarrow \text{bzac} \rightarrow \text{acac} \rightarrow \text{dipm}. \qquad \text{(Eq. C.10)}$$

This is evidently the order of the electron-attracting and releasing effects of the groups R_5 and R_6, as can be seen by comparing the Hammett's constants R of these groups:

	CF_3	C_6H_5	CH_3	$C(CH_3)_3$
R_{meta}	0.43	0.06	−0.07	−0.10
R_{para}	0.54	−0.01	−0.17	−0.197

Electron-attracting ⟷ electron-releasing

Spectral data have shown that the values of K increase with the increase in the number and size of the alkyl groups in the substituents R_1–R_4 on diam (cf. Fig. C.7). This is to be expected from the consideration of the steric effect. The observed order is [33]:

sym-dmen → unsym-dmen → sym-deen → tmen →
→ unsym-deen → teen (Eq. C.11)

Figure C.11a shows some relevant data. It can be noted that, among dmen's and deen's, the chelates of the sym- (or N,N′-) form are more susceptible to solvent attack than those of the unsym- (or N,N-) form. Figure C.11b shows how this difference comes about.

It can also be seen that, even in the case of teen, solvents with a high value of DN can form sizable amounts of octahedral species by forcing thier way through the bulky ethyl groups.

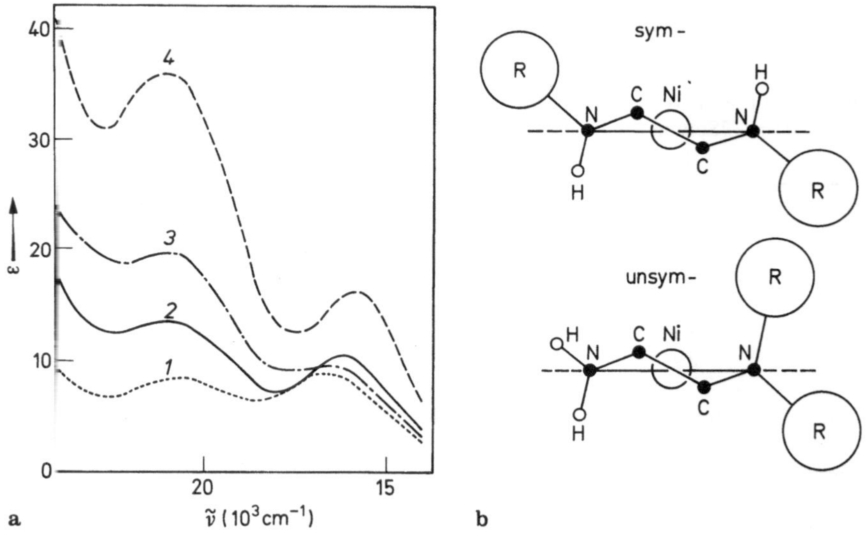

Fig. C-11 (a). Absorption spectra of [Ni(diam) (acac)] $B(C_6H_5)_4$ in acetone, where diam is: (*1*) sym-dmen; (*2*) unsym-dmen; (*3*) sym-deen; (*4*) unsym-deen. **(b)** Steric effects in the Ni-diamine rings as viewed along the chelate plane. The alkyl groups in sym-diamine chelates can occupy the equatorial sites, but half of them in unsym-diamine chelates occupy the axial sites, hindering the coordination of axial ligands (cf. Fig. C-3). (After Nga et al. [33])

C.IV.3 Nitrato- and Bis(β-diketonato)-Complexes

With the chelates of type III and IV (Fig. C.7), i.e., 6-coordinated bis(β-diketonato)- and nitrato-complexes, it was found that:
1) The solids and solutions of the type-IV chelates containing NO_3^- ion are always blue or green. Infrared studies of the solids have shown that NO_3^- acts as a chelating ligand, forming an octahedral entity [Ni(diam)(dike)NO_3]. In solvents of low DN, they dissolve without structural change to form non-conducting solutions. If the DN is higher, a part of the NO_3^- is displaced by solvent molecules and an equilibrium is established:

$$[\mathrm{Ni(diam)(dike)NO_3}] + 2\mathrm{Solv} \rightleftharpoons [\mathrm{Ni(diam)(dike)(Solv)_2}]^+ + \mathrm{NO_3}^- \qquad \text{(Eq. C.12)}$$

This equilibrium is only accompanied by a relatively slight spectral change [30–32].
2) In the type-III complexes containing an additional dike ligand, the octahedral molecule [Ni(diam)(dike)$_2$] is again blue or green, and highly stable. These complexes dissolve in various kinds of organic solvents without structural and spectral changes.

It is interesting to note that the type-III complexes are very soluble in highly non-polar solvents, e.g., CCl_4, benzene or hydrocarbons. Table C.6a lists some of their solubilities in these solvents; for example, [Ni(tmen)(acac)$_2$] is as soluble as 220 g/dm^3 solvent in n-C_7H_{16} [37].

It may be of further interest to compare these data with the solubilities of some metallic acetylacetonates in Table C.6b [38]. Here chelates of tervalent metals are highly soluble in C_6H_6, but much less soluble in n-C_7H_{16}. Chelates of the divalent metals are nearly insoluble in both solvents. This large difference in solubility is

Table C.6a. Solubilities (10^{-2} mol dm^{-3}, 20 °C) of some mixed chelates of Ni(II) in non-polar solvents. (After Saito et al. [37])

Chelate	C_6H_6	CCl_4	Heptane
[Ni(teen)(acac)$_2$]	100	140	55
[Ni(tmen)(acac)$_2$]	140	100	60
[Ni(tmen)(tfac)$_2$]	130	64	13
[Ni(tmen)(dipm)$_2$]	43	28	29

Table C.6b. Solubilities (g/100 g solvent, 20 °C) of some metallic acetylacetonates in non-polar solvents. (After Fernelius and Bryant [38]).

Chelate	C_6H_6	$C_6H_5CH_3$	Heptane
Al(acac)$_3$	35.9	16.0	0.1
Cr(acac)$_3$	30.6	11.6	0.05
Co(acac)$_3$	14.4	4.2	0.03
Fe(acac)$_3$	52.1	21.6	0.11
Co(acac)$_2$	0.6	–	–
Cu(acac)$_2$	0.1	0.06	0.01
Pd(acac)$_2$	1.3	0.66	0.02

certainly due to the fact that the central tervalent cations are effectively shielded by three organic ligands with six protruding methyl groups. As a result these chelates behave very much like non-polar organic molecules in solution. On the other hand, the central divalent cations and polar groups of the two ligands surrounding them are not sufficiently well shielded from outer influences, so that the intermolecular attraction in the solid is much stronger. In fact, this is the reason for the easy trimerization of $Ni(acac)_2$ described above. Similar polymerization is also known to occur with $Fe(acac)_2$, $Mn(acac)_2$, $Co(acac)_2$ and $Zn(acac)_2$.

It is, however, peculiar that each of these chelates is polymerized in its own way. $Fe(acac)_2$ is dimeric, $Co(acac)_2$ is tetrameric, while $Zn(acac)_2$ and $Mn(acac)_2$ are trimeric. Even the latter two are each of a different structure; $Zn(acac)_2$ contains one ZnO_6 unit and two ZnO_5 units, while $Mn(acac)_2$ contains one octahedral and two trigonal-prismatic MnO_6 units [38a, b]. The reason for these differences is still unknown.

$Cu(acac)_2$ and $Pd(acac)_2$ consist of monomers; however, the intermolecular attraction acting between their planar molecules can still be expected to be much stronger than that acting in tervalent metal chelates. The divalent metal chelates are therefore appreciably soluble only in solvents of high polarity.

The solubility data in Table C.6 clearly shows that the introduction of a bulky alkylated diamine molecule to the coordination sphere of $Ni(dike)_2$ renders the latter as non-polar as, or even more non-polar than, $[M(acac)_3]$, i.e., the intermolecular attraction between the chelates is extremely weakened. Correspondingly, the type-III chelates listed in this table generally have low melting points, even tending to easily vaporize upon heating slightly [37]. Most of these properties are also shared by the Co(II) and Zn(II) chelates of the corresponding structure [37a]. The diamine molecules gives these chelates, so to say, a new fin to swim in non-polar liquids and a new wing to fly in air.

C.IV.4 Halogeno- and Pseudohalogeno-Complexes

Chelates of the composition Ni(diam)(dike)X, where X^- = a halide or a pseudohalide ion [39–41], have also been synthesized.

The chelate Ni(tmen)(acac)NCS has been studied in detail by Hoshino et al. [41]. It dissolves in polar solvents (alcohols, acetone, DMSO etc.), forming solvated 6-coordinate species. However, non-conducting solutions are obtained with DCE. Dilute solutions in the latter yield spectra which are characteristically different from those of the 6- and 4-coordinate species. A comparison of these spectra with the spectra of other complexes studied by Sacconi et al. [42] led to the idea that a square pyramidal, 5-coordinate species [Ni(tmen)(acac)NCS] probably exists in such solutions. Increasing the concentration, or lowering the temperature, brings about a noticeable change in the spectrum. An analysis has shown that this is due to a shift in the equilibrium:

$$\underset{\text{monomer, 5-coordinated}}{2[\mathrm{Ni(tmen)(acac)NCS}]} \rightleftharpoons \underset{\text{dimer, 6-coordinated}}{[\mathrm{Ni(tmen)(acac)(NCS)}]_2}, \qquad \text{(Eq. C.13)}$$

to the right hand side with the formation of a dimeric species. The same dimeric structure also seems to exist in crystals of Ni(tmen)(acac)NCS. Infrared data [39, 41] indicate that the two Ni^{2+} ions in the dimer are linked by two -NCS- bridges.

It is well known that Ni^{2+}, like other cations of the first-row transition metals, prefers to combine with the N-end of NCS^- rather than the S-end. The bridges in the dimer will thus be such as:

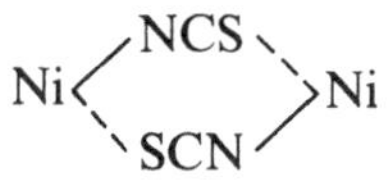

The weaker (Ni---SCN) bonds are easily broken even by a non-polar solvent, especially at higher temperatures. This seems to be the reason for the above equilibrium. It is interesting to note that the other end of the bridge (Ni—NCS) can also be cleaved in CH_3NO_2 (which is a little more polar than DCE). Hence, the spectrum observed in CH_3NO_2 clearly shows the existence of three species with 4, 5 and 6-coordinate structures, which again are in a thermochromic equilibrium:

	$[Ni(tmen)(acac)]^+$	[Ni(tmen)(acac)NCS]	Dimer
C.N.	4	5	6
Conc.	Low	⟷	High
Temp.	High	⟷	Low

(Eq. C.14)

In terms of DN, and AN to be discussed in the next chapter, we can summarize the results as follows. DCE(DN = 0, AN = 17) and CH_3NO_2(DN = 2.7, AN = 21) are both poor donors. Hence, they cannot effectively solvate the dissolved complexes. However, CH_3NO_2 can partially displace the NCS^- out of the coordination sphere of [Ni(tmen)(acac)NCS] due to its higher AN, i.e., stronger tendency to solvate anions. It is interesting to note that another interplay of DN and AN also exists among more polar solvents; according to a spectral titration study, the predominant species in acetone (DN = 17, AN = 13) seems to be [Ni(tmen)acac)(NCS)(Solv)], while $[Ni(tmen)(acac)(Solv)_2]^+$ is formed in DMSO (DN = 30, AN = 19). In the latter case the effect of solvent molecule is much stronger, both as a solvating donor and an ionizing acceptor. The weaker solvent, acetone, can only occupy the vacant site in the coordination sphere of the 5-coordinate complex. It cannot efficiently displace the NCS^- ion.

Different behaviors are found among the complexes Ni(tmen)(acac)X, where $X^- = Cl^-$, Br^- or I^- [41]. Their extensive ionization, and formation of $[Ni(tmen)(acac)(Solv)_2]^+$, also take place in highly polar solvents. In non-polar solvents the ionization of X^- is suppressed; instead, a different type of reaction involving a disproportionation of coordination numbers takes place in solution:

$$\underset{\text{C.N.}\quad 5}{2[Ni(tmen)(acac)X]} \rightarrow \underset{6}{[Ni(tmen)(acac)_2]} + \underset{4}{[Ni(tmen)X_2]} \qquad \text{(Eq. C.15)}$$

This reaction was first discovered, when crystals of these complex halides containing 1-$2H_2O$ per molecule were gently heated under reduced pressure [40]. The water molecules, which are coordinated in the original octahedral complex, are the first to be lost. The anhydrous complex is then decomposed according to (Eq. C.15); as noted above, the complex $[Ni(tmen)(acac)_2]$ is volatile, so it sublimes away and condenses as a blue sublimate on the outlet of the vessel, while $[Ni(tmen)X_2]$ remains at the bottom.

The latter complex was found [42] to be tetrahedral when $X^- = Br^-$ (violet) or I^-(green), both in the solid and in non-polar solvents. This rather unusual geometry for nickel(II) is governed by the bulkiness of tmen and the repulsion between the large halide ions. When $X^- = Cl^-$, the complex was found to be polymeric and octahedral (green) in the solid state. It is easily converted into the tetrahedral monomer in a non-polar solvent (e.g., o-$C_6H_4Cl_2$). Hence, a thermochromic change from green to violet is observed upon heating such solutions. These complexes thus present interesting chromotropic features by themselves.

The disproportionation (Eq. C.15) goes almost to completion when $X^- = Cl^-$. When $X^- = Br^-$, a small amount of red crystals (probably [Ni(tmen)(acac)]Br) is formed along with the violet complex [Ni(tmen)Br_2]. When $X^- = I^-$, all of the original complex is converted into red [Ni(tmen)(acac)]I when heated, and hardly any of the blue sublimate is formed. Therefore, the decomposition can be formulated as follows:

$$[\text{Ni(tmen)(acac)X}] \begin{array}{l} \nearrow \text{Disproportionation (a)} \\ \searrow [\text{Ni(tmen)(acac)}]^+ + X^- \text{ (b)} \end{array} \qquad \text{(Eq. C.16)}$$

The relative ease of these reactions is: (a) $\gg$ (b), (a) $>$ (b) and (a) $\ll$ (b) for $X^- = Cl^-$, Br^- and I^-, respectively.

When the hydrates of Ni(tmen)(acac)X.nH_2O(n = 1 or 2) are dissolved in DCE, acetone or CH_3NO_2, decompositions similar to (a) and (b) take place in the resulting solutions. Their spectra are fairly complex, but their general features indicate the occurence of such reactions, and their relative ease is comparable to that observed in the thermal decomposition of the solids. The results in DCE and acetone are quite similar. The tendency for ionization (b) increases remarkably in CH_3NO_2, corresponding to its higher AN, but otherwise the result is still similar.

Although the evidence is still not sufficient, we can probably assume that such a disproportionation takes place via a dimer, $[\text{(acac)(tmen)Ni}\langle{}^{X}_{X}\rangle\text{Ni(tmen)(acac)}]$, similar to that observed with NCS-complexes. The ease of such a reaction should depend on the ease of formation of such a dimer and its high deformability in the course of the reaction. Thus, it should be promoted by Cl^- and Br^- ions, since they can form reasonably strong ionic Ni-X-Ni bridges that can still be bent and broken at either end. On the other hand, I^- is too bulky and weakly bound to Ni^{2+}, so that it merely tends to ionize off. NCS^- ions can form a dimer with long bridges, whose N-ends are strongly bound but whose S-ends are only weakly bound. This hinders such a disproportination. At any rate, the nature of these complexes and their reactions are worth further studies.

C.V Chelates of Schiff Bases and Related Ligands: Tetrahedral-Square Planar Equilibria

In addition to the octahedral-square planar equilibria discussed above, there are also thermochromic equilibria involving tetrahedral Ni(II) complexes. Some of these have already been referred to in Chapter B. As in the case of Co(II), tetrahedral complexes are formed when the Ni-L bond is rather weak and ionic, as in $[NiCl_4]^{2-}$ or $[NiBr_4]^{2-}$, and/or the ligand is very bulky, as in $[Ni(HMPA)_4]^{2+}$, where the interligand repulsion favors a tetrahedral geometry. However, as noted before (Chaps. B.VI and B.VII), Ni(II) is much more reluctant to form such complexes than Co(II) under the same conditions.

Comparing the electron distributions of Co^{2+} and Ni^{2+} in octahedral (Oh) and tetrahedral (Td) complexes, one can note that an Oh complex of Ni^{2+} ($t_{2g}{}^6e_g{}^2$) is more stable than that of Co^{2+} ($t_{2g}{}^5e_g{}^2$), since it contains an additional electron in the low-lying t_{2g} orbital. A Td complex of Ni^{2+} ($e^4t_2{}^4$), on the other hand, is less stable than that of Co^{2+} ($e^4t_2{}^3$), since it possesses an additional electron in the high-lying t_2 orbital. Thus, the conversion of a complex from Oh to Td is expected to be much more difficult in the case of Ni^{2+}, as has been observed in so many instances.

Other examples of such equilibria are found in the works of Sacconi, Eaton, Holm and their collaborators, where the problem was studied extensively using magnetic, spectral and NMR techniques. Since good reviews are available [44, 45–48], a brief outline of Sacconi's results on a few Schiff base chelates of nickel(II), i.e., bis-(salicylaldiminato)nickel(II) derivatives, will be given in some detail as a representative example of such equilibria.

These chelates are of the following general structure:

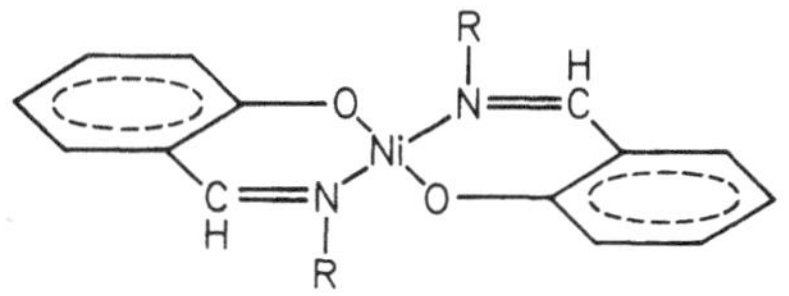

When R is equal to methyl, the crystals that are obtained are green and diamagnetic and consist of trans-planar molecules as shown above. In non-polar solvents such as C_6H_6 or $CHCl_3$, however, the complex is slightly paramagnetic ($\mu_{eff} = 1–2$BM). The magnitude of this paramagnetism depends on the nature of the solvent, as well as the concentration and temperature. This seems to be due to the formation of polymerized species. In fact, by heating the solid crystals up to 180 °C, they are converted into a yellow modification that shows a typical octahedral spectrum. This material can be formulated in terms of the infinite polymer shown in Fig. C.12 and taken to be the limiting form of the polymerization observed in solution.

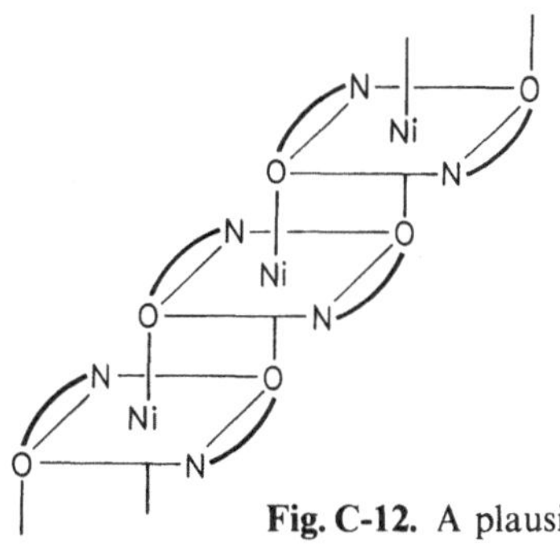

Fig. C-12. A plausible structure of polymeric [Ni(N-methyl-salicylaldimine)$_2$]

As R gets larger, and especially bulkier, the tendency for such a polymerization decreases. Instead another kind of structural change becomes evident. When R is equal to ethyl or n-propyl, the chelates are green, square planar and diamagnetic. When R is equal to isopropyl or tert-butyl the solid becomes brown and definitely paramagnetic (μ_{eff}: 3 BM). Their solutions in non-polar solvents exhibit a peculiar thermochromic change, which indicates that two forms of the same chelate, a "green form" and a "brown form", exist in thermal equilibrium. In the case of the isopropyl chelate, the amount of the brown form increases at higher temperatures, while the reverse is true for the tert-butyl chelate.

Sacconi et al. explained these phenomena in the following way. When R is a bulky group, such as tert-butyl, the coordination of the oxygen of the neighboring ligand is sterically hindered, so that the two ligands are twisted to a tetrahedral arrangement. However, this must accompany a drastic change in the electronic state, with a considerable weakening of the ligand field strength (or coordinate bonds). The chelate thus becomes paramagnetic (high-spin), exhibiting a color which is quite distinct from that of the planar methyl or ethyl chelate.

In a non-polar solvent, a fraction of the tetrahedral, tert-butyl chelate molecules will be twisted back to the energy-rich planar form by thermal movements. The actual percentage increases with increasing temperature. The peculiar thermochromism shown below is the result.

$$\underset{\text{"brown form"}}{\text{tetrahedral}} \underset{\text{Cool}}{\overset{\text{Heat}}{\rightleftarrows}} \underset{\text{"green form"}}{\text{square planar}} \qquad \text{(Eq. C.17)}$$

In the case of the isopropyl chelate, the substituent is not as bulky as tert-butyl. Thus, the interligand repulsion is smaller and, although the solid is composed of tetrahedral chelates, the dissolved particles tend to be square planar in non-polar solvents when the temperature is low. Thermal movements will oppose this tendency, twisting the chelate into a slightly energy-rich tetrahedral form. The resulting thermochromic change takes place in the opposite direction.

Sacconi et al. also studied many other related chelates, containing long chain alkyls and aryls. They studied the effects of the various substituent groups on R and the benzene ring of the original ligand on these equilibria. Although there often seems to exist a delicate balance between steric and electronic effects, they could explain and classify their results skillfully.

Here the reader may have been puzzled to learn that "the square planar chelates are green, the polymeric 6-coordinate one is yellow and the tetrahedral ones are brown". Certainly their colors are very different from the colors of common Ni(II) complexes of the same geometries. These colors are seemingly due to the peculiar patterns of their spectra, including strong ligand and CT bands in the UV and blue parts which tail off into the longer wavelength region. Careful assignments of the bands, however, justify Sacconi's viewpoints.

Another example of tetrahedral-square planar equilibria of Ni(II) chelates can be found in the studies of aminotropone iminato chelates.

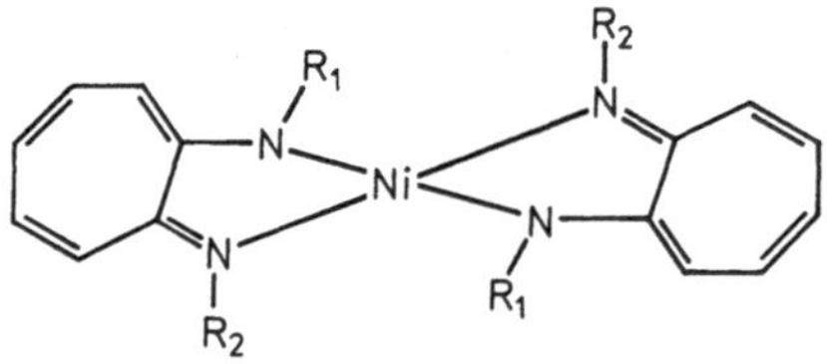

An interplay between electronic and steric effects also takes place. In this case NMR techniques have been very successfully applied to elucidate the situation [45, 46]. A collection of NMR and thermodynamic data on these and other planar-tetrahedral equilibria can be found in the reviews by Holm and O'Connor [47, 48].

In concluding this section, it may be added that most of these equilibria were observed in non-polar or non-coordinating solvents (e.g., C_6H_6 and $CHCl_3$, or less volatile ones such as xylene and bibenzyl). Spectral studies on the melts of the chelates themselves were also made in certain cases. In more polar solvents, the situation is complicated further by solvated species.

C.VI Thermochromic Systems Involving Formation of Five-Coordinate Complexes and Linkage Isomerization

C.VI.1 Formation of Five-Coordinate Complexes

At the end of Chap. C.IV and also earlier (cf. $[Ni(bapp)]^{2+}$, C.II), mention was made of the formation of 5-coordinate complexes of nickel(II). Thermochromic equilibria involving such complexes are seemingly rare. They may, however, be more common than is usually expected. A few more examples will be presented here.

An interesting example is the chelate $[Ni(dacoda)]$, studied by Averill et al. [49] and then by Billo [50]. The ligand dacoda^{2-} forms an orange, square planar nickel(II)

$$^{\ominus}OOC—CH_2—N \qquad N—CH_2—COO^{\ominus}$$

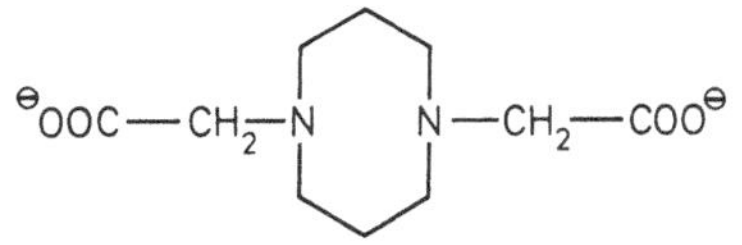

complex $[Ni(dacoda)]$ and a bluish green monohydrate $[Ni(dacoda)(H_2O)]$. The latter is presumably of the structure shown in Fig. C.13a.

Aqueous solutions of these two compounds are of the same bluish green color. Their absorption spectra are nearly of the same shape as that of the solid monohydrate, which is remarkably different from that of a common octahedral complex of nickel(II) (cf. Fig. C.13b; note that edda, $(CH_2NHCH_2COO^-)_2$, forms a typical octahedral complex, $[Ni(edda)(H_2O)_2]$, in aqueous solution). Averill et al. assumed that the steric hindrance of the $(CH_2—CH_2—CH_2)$-groups protruding above and below the plane of the chelate allows room for only one molecule of water in the coordination sphere, so that a 5-coordinate species $[Ni(dacoda)(H_2O)]$ is formed in solution as in the solid phase.

Billo found this solution to be thermochromic, becoming strongly yellowish upon heating. A noticeable inert salt effect also causes the same color change, as shown in Fig. C.13c. He concluded that the water molecule, coordinated only weakly in comparison with the hydrophobic ligand, easily leaves its site. He also determined some thermodynamic parameters of this desolvation equilibrium.

Recently Fukuda et al. [51] prepared similar chelates with the derivatives of "dachda" shown below:

$$^{\ominus}OOC—CHR—N \qquad N—CHR—COO^{\ominus}$$

R = H : dachda$^{2\ominus}$
R = Me : dachdma$^{2\ominus}$
R = Et : dachdea$^{2\ominus}$

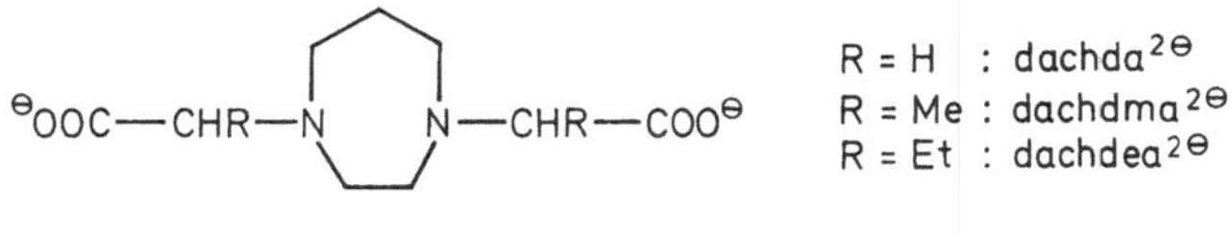

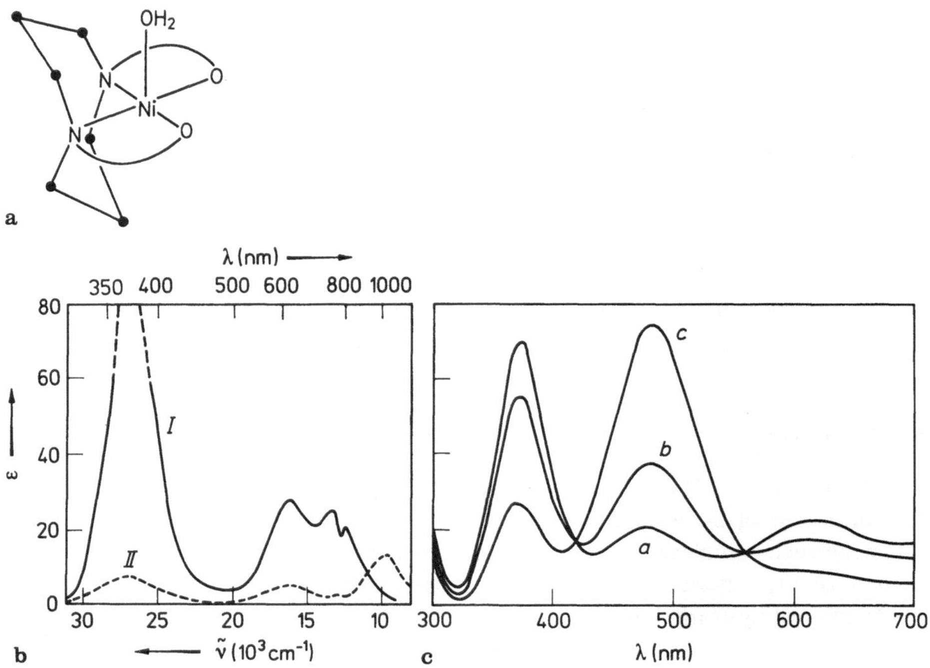

a b c

Fig. C-13 (a). Plausible structure of [Ni(dacoda) (H_2O)]. **(b):** Absorption spectra of [Ni(dacoda) (H_2O)] (I) and [Ni(edda) $(H_2O)_2$] (II) in aqueous solutions. **(c):** Effect of conc. $NaClO_4$ on the spectrum of [Ni(dacoda) (H_2O)] (Conc.: 2.12×10^{-3} M; 25 °C). $NaClO_4$ concentration: *a*, 4 M; *b*, 6 M; *c*, 7.2 M. (After Averill et al. [49] and Billo [50])

By diminishing the size of the central ring of dacoda^{2-}, axial coordination of a water molecule is expected to be favored. Surprisingly the situation is not so simple; the chelates [Ni(dachdma)] and [Ni(dachdea)] tend to be more planar in solution than [Ni(dacoda)] [50]. The ease of formation of the 5-coordinate species decreases, approximately, in the order of dacoda $>$ dachda $\gg$ dachdma $\gtrsim$ dahadea. The steric effect hindering axial coordination of a water molecule in these chelates is not only due to the $(CH_2)_n$-bridges above and below the plane of the chelate, but also due to the R groups which protrude outside of the chelate ring and produce a hydrophobic atmosphere about the entire complex. Comparative studies on these and related chelates are now being carried out.

C.VI.2 Isomerization Equilibria: The Nitro-Nitrito Equilibria

In the foregoing sections we have treated, for the most part, thermochromic or solvatochromic equilibria involving species of different coordination numbers. Such equilibria can also occur when some of the ligands (A) in a fixed coordination geometry are reversibly substituted by others (B) which are far apart from (A) in the spectrochemical series.

An interesting example is the unique thermochromism of the $[NiL_4(NO_2/ONO)_2]$-type complexes in which L is an amine or L_2 is a diamine [52]. All of these complexes are typically octahedral, exhibit three d-d bands and are paramagnetic (μ_{eff}: ca. 3 B. M.) (cf. Chap. C.II). However, some of them are red while others are blue, exhibiting d-d bands and infrared bands at different positions. Goodgame and Hitchmann [52] ascribed the red ones to nitro complexes involving Ni—NO_2 bonds, and the blue ones to nitrito complexes involving Ni—ONO bonds. [$Ni(NH_3)_4(NO_2)_2$], [$Ni(en)_2(NO_2)_2$] and [Ni(N-methyl-en)$_2$(NO_2)$_2$] belong to the former type of complex, while [$Ni(py)_4(ONO)_2$], [Ni(unsym-dmen)$_2$(ONO)$_2$] and [Ni(sym-deen)$_2$(ONO)$_2$] belong to the latter. In addition they found that the chloroform solutions of the last two chelates show two bands at ca. 500 and 590 nm, which are due to the nitro- and nitrito-isomers of the same complex. The solutions are remarkably thermochromic (cf. Fig. C.14) due to a shift in the folllowing equilibrium:

$$\underset{\text{red}}{[Ni(diam)_2(NO_2)_2]} \rightleftarrows \underset{\text{blue}}{[Ni(diam)_2(ONO)_2]}. \qquad \text{(Eq. C.18)}$$

The value of ΔH for this reaction was calculated using the data given in Fig. C.14 and was found to be $-9.6\,kJ\,mol^{-1}$.

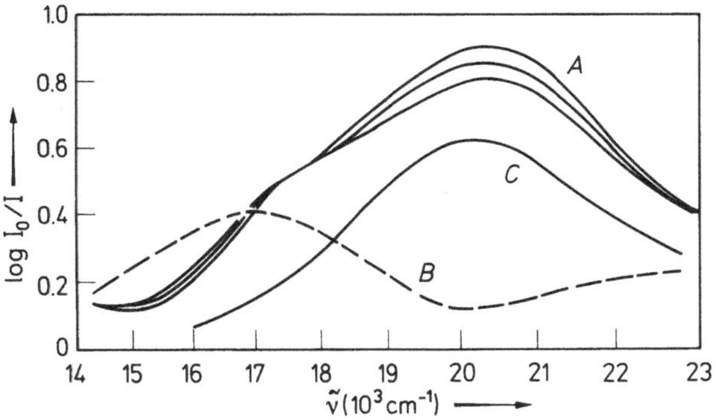

Fig. C-14. Absorption spectra of: *A*, [Ni(sym-deen)$_2$ (ONO)$_2$] in $CHCl_3$, 1.5×10^{-2} M, temp. (from top to bottom): 19, 32, and 44 °C; *B*, [Ni(sym-deen)$_2$ (ONO)$_2$], solid reflectance; *C* [Ni(N-ethyl-en)$_2$ (NO_2)$_2$] in $CHCl_3$, 8.15×10^{-3} M. Cell thickness 5 cm. (After Goodgame and Hitchman [52])

All of these phenomena are evidently due to the steric effect. It is easy to see that the group —ONO (—O\N/O) is less bulky than —NO_2 (—N<O/O) in the coordination sphere of Ni(II) [53]. When NO_2^- ions are coordinated above and below the plane of [$Ni(NH_3)_4$]$^{2+}$ or [$Ni(en)_2$]$^{2+}$, they tend to be bound with their nitrogen end. In doing so, they can exert a stronger ligand field, as can be seen from the spectrochemical series. Bulky substituents above and below the plane of the chelate cause steric repulsion and hinder the NO_2^- ions to be bound in this way. Hence, they tend here to be bound as —ONO with less steric requirements. The decrease in the ligand field strength accompanying this structural change (isomerization) leads to a change in color from red to blue.

When these two tendencies are nearly balanced, it is possible that solute-solvent interactions and thermal movements in a solution of a nitrito complex in a suitable solvent make the turnover from —ONO to —NO_2 possible, changing the color from blue to red. On heating such a solution, the energy-rich nitrito complex is again formed, so that the solution becomes bluish. This is what Fig. C.14 reveals.

Such a turnover is also observed with cobalt(III) complexes, such as $[Co(en)_2(ONO)_2]^+$. The color of an aqueous solution changes gradually from pink to yellow. Spectral changes with two isosbestic points take place corresponding to the two-step reaction [54]:

$$[Co(en)_2(ONO)_2]^+ \rightarrow [Co(en)_2(NO_2)(ONO)]^+ \rightarrow [Co(en)_2(NO_2)_2]^+ \qquad \text{(Eq. C.19)}$$

Kinetic studies have shown that this reaction proceeds via an intramolecular mechanism, i.e., via a turnover of NO_2^- in the coordination sphere. This reaction, however, is irreversible; it even occurs in the solid state. In contrast to Co^{3+}, Cr^{3+} only forms —ONO complexes, such as $[Cr(NH_3)_4(ONO)_2]^+$ and $[Cr(en)_2(ONO)_2]^+$, with no tendency to isomerize into —NO_2 complexes. The dynamic changes observed among nickel(II) chelates are thus quite peculiar.

References

1. Gutmann, V.: "Coordination Chemistry in Non-Aqueous Solutions", Springer-Verlag, Wien (1968)
2. Gutmann, V.: "The Donor-Acceptor Approach to Molecular Interactions", Plenum, New York (1978)
3. Cotton, F. A., Wilkinson, G.: "Advanced Inorganic Chemistry", 4th Ed., Wiley, New York (1980), p. 787
4. Basolo, F. et al.: J. Am. Chem. Soc. **76**, 956 (1954)
5. Jørgensen, C. K.: Acta Chem. Scand. **11**, 399 (1957)
6. Sone, K., Kato, M.: Z. anorg. allg. Chem. **301**, 277 (1959)
7. Leussing, D. L. et al.: J. Phys. Chem. **66**, 1544 (1962)
8. Farago, M. E. et al.: J. Chem. Soc. **1968A**, 48
9. Nyburg, S. C. et al.: Proc. Chem. Soc. **1961**, 297
10. Jørgensen, C. K.: "Inorganic Complexes", Academic Press, London (1963), p. 100
11. Higginson, W. C. E. et al.: Inorg. Chem. **3**, 463 (1964)
12. Nyburg, S. C., Wood, J. S.: Inorg. Chem. **3**, 468 (1964)
13. Gillard, R. D., Sutton, H. M.: J. Chem. Soc. **1970A**, 1309
14. Gillard, R. D., Sutton, H. M.: ibid. **1970A**, 2172
15. Gillard, R. D., Sutton, H. M.: ibid. **1970A**, 2175
16. Lifschitz, I. et al.: Z. anorg. allg. Chem. **242**, 97 (1939)
17. Lifschitz, I., Bos, T. G.: Rec. Trav. Chim. **59**, 407 (1940)
18. Lifschitz, I., Dijkema, K. M.: ibid. **60**, 581 (1941)
19. Curtis, N. F., House, D. A.: J. Chem. Soc. **1965**, 6194
20. Iwasaki, N. et al.: Z. anorg. allg. Chem. **412**, 170 (1975)
21. Gibson, J. G., McKenzie, E. D.: J. Chem. Soc. **1971 (A)**, 1029
22. Mochizuki, K. et al: Bull. Chem. Soc. Jpn. **53**, 2535 (1980)
23. Anichini, A. et al.: Inorg. Chim. Acta **24**, L21 (1977)
24. Sabatini, L., Fabbrizzi, L.: Inorg. Chem. **18**, 438 (1979)
25. Cotton, F. A., Fackler, J. P. Jr.: J. Am. Chem. Soc. **83**, 2818 (1961)
26. Fackler, J. P. Jr., Cotton, F. A.: ibid. **83**, 3775 (1961)
27. Bullen, G. J. et al.: Nature **189**, 291 (1961)
28. Yoshida, I. et al.: Bull. Chem. Soc. Jpn. **45**, 1411 (1972)
29. Chamberlain, C. S., Drago, R. S.: Inorg. Chim. Acta **32**, 75 (1979)

30. Fukuda, Y., Sone, K.: J. Inorg. Nucl. Chem. **34**, 2315 (1972)
31. Fukuda, Y., Sone, K.: ibid. **37**, 455 (1975)
32. Fukuda, Y. et al: Bull. Chem. Soc. Jpn. **49**, 1017 (1976)
33. Nga, N.T. et al.: ibid. **50**, 154 (1977)
34. Mayer, U.: "Ions and Molecules in Solutions", (Ed. Tanaka, N. et al.) Elsevier, Amsterdam (1983), p. 219
35. Soukup, R.W.: Chemie in unserer Zeit **17**, 163 (1983)
36. Ogino, H. et al.: "Abstracts of VI International Symposium on Solute-Solute-Solvent Interactions", Minoo (Japan), No. 5P-30 (1982)
37. Saito, Y. et al: Bull. Chem. Soc. Jpn. **54**, 196 (1981)
37a. Shika, Y.: Unpublished data (1984)
38. Fernelius, W.C., Bryant, B.E.: "Inorganic Synthesis", Vol. 5, (Ed. Moeller, Th.) McGraw-Hill, New York (1957), p. 111
38a. Bennett, M.J. et al.: Nature **213**, 174 (1967)
38b. Shibata, S. et al.: Chem. Lett. **1984**, 485
39. Hoshino, N. et al.: Transition Metal Chem. **4**, 183 (1979)
40. Hoshino, N. et al.: Chem. Lett. **1979**, 437
41. Hoshino, N. et al.: Bull. Chem. Soc. Jpn. **54**, 420 (1981)
42. Sacconi, L. et al.: Inorg. Chem. **4**, 943 (1965)
43. Sacconi, L. et al.: Inorg. Chem. **6**, 262 (1967)
44. Sacconi, L.: "Transition Metal Chemistry", Vol. 4, (Ed. Carlin, R.L.) Marcel Dekker, New York (1968), p. 199
45. Eaton, D.R. et al.: J. Am. Chem. Soc. **85**, 397 (1963)
46. Eaton, D.R., Phillips, W.D.: J. Chem. Phys. **43**, 392 (1965)
47. Holm, R.H.: Acc. Chem. Res. **2**, 307 (1969)
48. Holm, R.H.: O'Connor, M.J.: Prog. Inorg. Chem. **14**, 408 (1971)
49. Averill, D.F. et al.: Inorg. Chem. **11**, 2344 (1972)
50. Billo, E.J.: Inorg. Chim. Acta **37**, L533 (1979)
51. Fukuda, Y. et al.: Chem. Lett. **1984**, 1309
52. Goodgame, D.L.M., Hitchman, N.A.: Inorg. Chem. **5**, 1303 (1966)
53. Huheey, J.E.: "Inorganic Chemistry", 2nd ed., Haper & Row, New York (1978), p. 473
54. Miyoshi, K. et al.: Inorg. Chem. **22**, 1839 (1983)

CHAPTER D

Chromotropic Phenomena of Copper(II) Chelates

D.I Introduction: Structure and Spectra of Copper(II) Chelates

A large number of chromotropic phenomena are also observed among the chelates of copper(II). We shall now compare their structures with those of the Ni(II) complexes discussed in Chapter C in order to illustrate certain points of interest.

A number of texts and reviews, both elementary and advanced [1–5], are available on this topic and are recommended to the interested reader.

Figure C.1 shows that when an octahedral (Oh), high-spin complex of Ni(II) is converted into a square planar (SP), low spin one due to an increase in equatorial ligand field strength, the lone electron in the $d_{x^2-y^2}$ orbital of the Oh complex is forced by the four equatorial ligands into the d_{z^2} orbital to form an electron pair. This pair, in turn, strongly hinders the approach of axial ligands, so that they are driven out. The result is a square planar structure. If the equatorial ligand field strength is strong enough, the loss of the axial Ni-L bonds is overcome by a strengthening of the equatorial bonds cuased by the removal of the odd electron in $d_{x^2-y^2}$. If the ligands possess vacant π-orbitals, additional π-bond formation (back donation) can further stabilize the square planar configuration.

In going from Ni^{2+} (d^8) to Cu^{2+} (d^9), where we have one more electron, the situation changes somewhat. In most complexes of Cu(II), it is known that the $d_{x^2-y^2}$ and d_{z^2} orbitals are occupied by a single electron and an electron pair, respectively. Hence, the four equatorial ligands of a Cu^{2+} ion surrounded by an octahedral field of ligands are more strongly bound to the Cu^{2+} ion than the axial ligands. This is because the former are only repelled by an odd electron in $d_{x^2-y^2}$, while the latter are repelled by an electron pair in d_{z^2}. Thus, in most Cu^{2+} complexes an axially elongated octahedral structure, which lies somewhere between a pure Oh structure and a pure SP one, is formed (Jahn-Teller effect).

We can now compare the energy diagrams of these systems. In Fig. C.1, the one for an Oh complex of Ni(II) is too common; it only shows how the five d-orbitals are split in an octahedral crystal field, and how the eight electrons are distributed. The one for a SP complex of Ni(II) shows that the $d_{x^2-y^2}$ orbital is much higher in energy than the remaining orbitals, owing to the repulsion from the equatorial ligands. Hence only the $d_{x^2-y^2}$ orbital remains empty, while the other orbitals are occupied by electron pairs.

The diagram for most Cu(II) complexes, shown in Fig. D.1B, again lies between these two extremes. Here the $d_{x^2-y^2}$ orbital is singly occupied, and strongly repelled by the firmly bound equatorial ligands. It is of the highest energy, followed by the d_{z^2}

orbital and the remaining three orbitals not facing the ligands. Of the latter orbitals, the d_{xy} orbital is of the highest energy because it lies in the XY plane and is repelled by the equatorial ligands somewhat more strongly than are the d_{xz} and d_{yz} orbitals.

Many chelates of Cu(II), such as $[Cu(en)_2]^{2+}$ and $[Cu(acac)_2]$, are of essentially planar structure in the solid. Their axial positions are sometimes unoccupied. More usually, they are occupied by counter anions, water (or other solvent) molecules of crystallization, or polar groups on other ligands, loosely bound to the central Cu^{2+} ion. In solution, these positions are often occupied by solvent molecules, which solvate the chelate from above and below its plane. In all of these cases, therefore, the equatorial and axial ligands are different. The former are usualy bound to the Cu^{2+} ion much stronger. If we change the axial ligands for a fixed set of equatorial ligands, the energy diagram will change as shown in Fig. D.1.

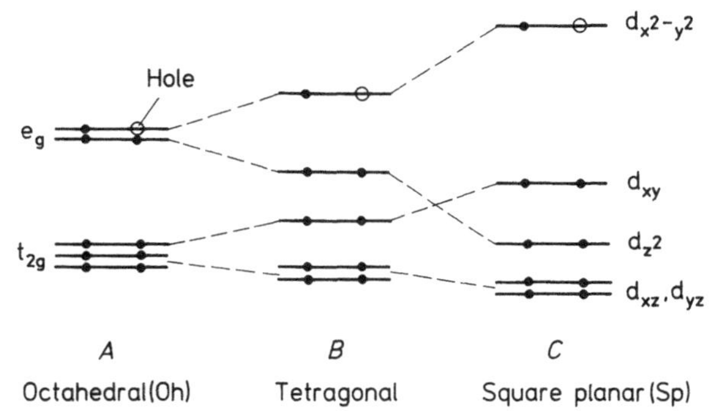

Fig. D-1. Energy diagram of Cu^{2+} in various ligand fields. Energy scale is only qualitative

When the ligand field strength of the axial ligands is equal to that of the equatorial ligands, the chelate will be nearly octahedral, i.e., its energy diagram will be similar to that of an Oh complex of Ni(II). On the other hand, for a very weak field strength of the axial ligands the chelate will be nearly planar. Its energy diagram will be similar to that of an SP complex of Ni(II). In both cases, an additional electron must be added to the highest orbital to convert the diagram from Ni(II) to Cu(II). Figure D.1 shows how the relative energy of each orbital changes when the ligand field strength of the axial ligands decreases from left to right. Therefore, a chelate such as $[Cu(AA)_2]$ (AA = bidentate ligand) can assume a series of structures with various "tetragonalities", or degrees of elongation from a regular octahedron, according to the coordinating abilities of the coexisting species (counter ions, solvent molecules, etc.).

The electronic spectrum of a Cu(II) complex is apparently simple. It only shows one broad d-d band in the visible region, corresponding to the excitation of an electron to the "hole" in $d_{x^2-y^2}$ from the remaining four orbitals. In general, the position ($\tilde{\nu}_{max}$) of this band is shifted to the blue with an increase in the ligand field strength of the equatorial ligands and a fixed set of axial ligands. Vice versa, this band is shifted to the red with an increase in the ligand field strength of the axial ligands and a fixed set of equatorial ligands.

These seemingly conflicting effects can again be understood using Fig. D.1. In the former case, the repulsion between the equatorial ligands and the electron in $d_{x^2-y^2}$ increases, so that this electron will become more and more energy-rich and unstable.

Unlike in Ni(II), it cannot escape into the d_{z^2} orbital which is now fully occupied, so that it must hold its position by any means. It is easy to see that no other electron will tend to fill the hole in the same orbital where the ligand repulsion is so strong. Thus, more energy is required for the excitation, shifting the $\tilde{\nu}_{max}$ to the blue in the order of increasing ligand field strength of the equatorial ligands. On the other hand, in the latter case, the electron pair in d_{z^2} (and, to a lesser extent, those in d_{xz} and d_{yz}) is repelled and instabilized more strongly by the axial ligands of higher ligand field strength. As a result less energy is needed to excite electrons to the hole in $d_{x^2-y^2}$, which shifts $\tilde{\nu}_{max}$ to the red.

The results of these two effects are brilliant changes in color from red to violet, blue and green for Cu(II) complexes. The values of λ_{max} range from ca. 500 nm to the near IR, depending on the changes in the equatorial and axial ligands. Most of these colors, and the chromotropic phenomena in the following sections, can be reasonably understood in terms of this picture.

D.II Acetylacetonates and Tetrammine-Type Complexes

D.II.1 Acetylacetonates: Solvatochromism, Thermochromism and Gas-Phase Spectra

In studying the chelates of Cu(II), we very often observe that they are notably solvatochromic. For example, the violet chelate [Cu(acac)$_2$] yields violet solutions in solvents of low polarity, which become more and more bluish as the polarity of the solvent increases.

Figure D.2 shows the spectra of this chelate in various solvents recorded by Belford et al. [6, 7], whose pioneering studies opened the gateway to an understanding of the spectra of Cu(II) complexes. Comparing the curves from top to bottom, we readily recognize that the d-d band is shifted remarkably to the red in going from $CHCl_3$ to piperidine. The molecule of [Cu(acac)$_2$] is known to be square planar, both in the crystal [8] and in gaseous state (this chelate sublimes easily in vacuum, enabling electron diffraction studies to be carried out [9]). In solution, we can therefore assume that two solvent molecules are loosely bound to the Cu(II) ion in the axial positions. The ligand field strength of these axial ligands increases in the order of their coordination ability, which, in turn, can be expressed in terms of DN (cf. Chap. C.IV). Although DN values for most of the solvents in Fig. D.2 are still unknown, a comparison with the DN of similar solvents (Table C.4) reveals that they increase in the order of $CHCl_3 \rightarrow$ dioxane $\rightarrow$ AmOH $\rightarrow$ py $\rightarrow$ piperidine. So the d-d band is shifted to lower frequencies in this order.

One also finds that the band in $CHCl_3$ or dioxane is split. The unsymmetrical shapes of the band in other solvents also indicate that they are composed or more than one band. In fact, Belford et al. analyzed these curves and found that all of them can be considered to be a superposition of three symmetrical bands, corresponding to the transitions $(d_{xz}, d_{yz}) \rightarrow d_{x^2-y^2}$, $d_{xy} \rightarrow d_{x^2-y^2}$, and $d_{z^2} \rightarrow d_{x^2-y^2}$, respectively. They could even explain the shift of each of these bands in a satisfactory way.

Although the general features of these solvatochromic data can be understood in terms of a (square planar-elongated octahedral) change, another structural change may

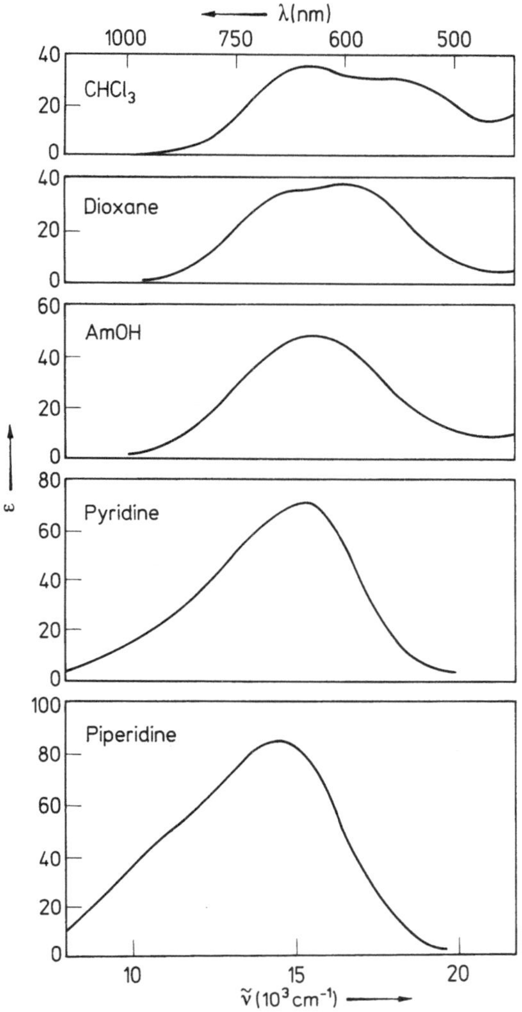

Fig. D-2. Absorption spectra of $[Cu(acac)_2]$ in various solvents. (After Belford et al. [7])

come into play when the ligand field strength of the axial ligands is sufficiently high. In such a case, the repulsion between an axial ligand and the electron pair in d_{z^2} is very strong, so that the electron pair tends to avoid the region about the ligand, if it is at all possible. This can be accomplished, for example, by hybridizing this orbital with the 4s and 4p orbitals. The electron pair can then occupy one of the hybrid orbitals, which points away from the ligand, leaving the other hybrid orbital directed towards the ligand vacant. The situation is shown schematically in Fig. D.3.

The axial Cu-L bond, strengthened by the removal of the interfering electron pair, pulls the Cu^{2+} ion out of the plane of the chelate to yield a square pyramidal structure. The coordination of another ligand L below the plane is now difficult, due to the strong repulsion between it and the displaced electron pair. The chelate rings, now bent downward, also hinder such an approach. The solvated complex formed in such a case assumes a 5-coordinate structure, $[Cu(acac)_2L]$.

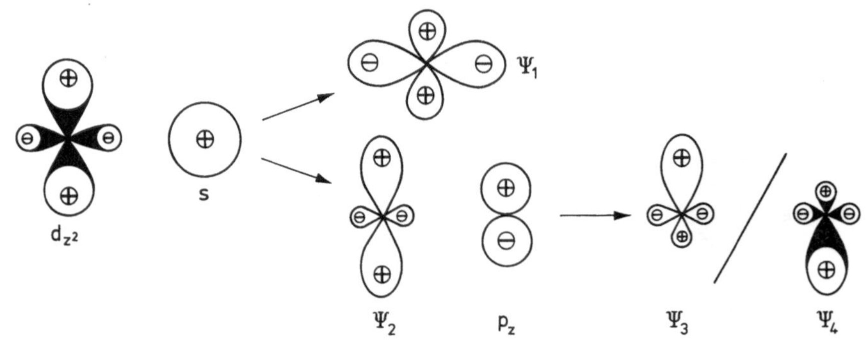

Fig. D-3. Hybridization of d_{z^2} with s yields Ψ_1 and Ψ_2. Another hybridization of Ψ_2 with p_z yields Ψ_3 and Ψ_4. The original d_{z^2}, s and p_z AO's are converted into three hybrid orbitals Ψ_1, Ψ_3, and Ψ_4. If an electron pair is originally in d_{z^2}, it can be driven out into Ψ_4 by the attack of a strong ligand from above the xy plane, to minimize the repulsion with that ligand (the shadows of d_{z^2} and Ψ_4 indicate this). (After Huheey [1])

In fact, such a case was observed by Graddon [10]. He found that, when pyridine is added, little by little, to a chloroform solution of [Cu(acac)$_2$], the observed spectral change indicates a shift in a single equilibrium,

$$[Cu(acac)_2] + py \rightarrow [Cu(acac)_2py], \qquad (Eq. D.1)$$

to the right hand side. When a large excess of pyridine is present, the spectrum of the resulting solution nearly coincides with that of a solution of [Cu(acac)$_2$] in pure pyridine. Similar results were also obtained with piperidine. These results show that the formation of a 5-coordinate species as depicted above really does take place, when the ligand field strength or coordination ability of the solvent (acting as the axial ligand) is very high (note that the DN values of pyridine and piperidine are by far much higher than those of other solvents in Fig. D.3).

In this connection, the results of Yokoi may be of interest. He found that a pyridine solution of [Cu(acac)$_2$] and also those of [Cu(bzac)$_2$] and [Cu(dipm)$_2$] are somewhat thermochromic [11]. The spectral changes observed between 0 °C and 60 °C are rather small, but the existence of isosbestic points suggests that there are two types of complexes in thermal equilibrium. Yokoi assumed that both are 5-coordinate [Cu(acac)$_2$py] species, with different ligand field strengths and different degrees of rotational freedom of the coordinated pyridine.

Another thermochromic study on [Cu(acac)$_2$] was carried out by Mizutani et al. [12]. They dissolved the chelate in two high-boiling solvents, $CHCl_2CHCl_2$(TCE) and n-octanol, respectively, and studied the spectra of the solutions up to 120 °C. The spectrum in TCE at room temperature was similar to that in $CHCl_3$ (cf. Fig. D.3). The shape of the d-d band did not change noticeably upon heating, except for a gradual increase in intensity. On the other hand, in n-octanol the shape of the d-d band changes remarkably upon heating. A broad shoulder appears on the blue side of the maximum (Fig. D.4a) This change suggests that the coordinate bonds between the OH groups of n-octanol and Cu(II) are broken, or notably weakened at 120 °C, so that the axial environments of the dissolved chelate become quite nonpolar.

Mizutani et al. also tried to measure the spectrum of [Cu(acac)$_2$] in the gaseous state, in order to determine the shape of the d-d band when the axial positions are

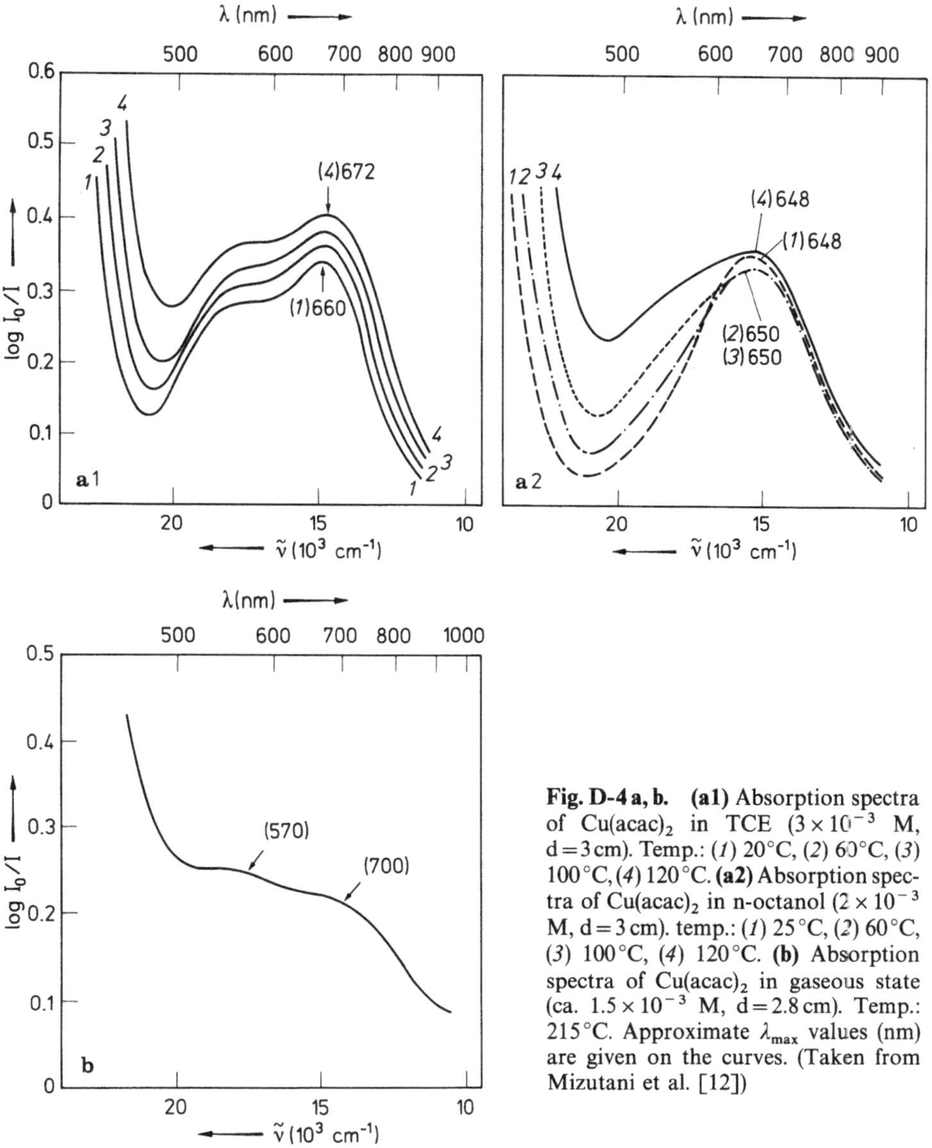

Fig. D-4 a, b. (a1) Absorption spectra of $Cu(acac)_2$ in TCE (3×10^{-3} M, d = 3 cm). Temp.: (*1*) 20 °C, (*2*) 60 °C, (*3*) 100 °C, (*4*) 120 °C. **(a2)** Absorption spectra of $Cu(acac)_2$ in n-octanol (2×10^{-3} M, d = 3 cm). temp.: (*1*) 25 °C, (*2*) 60 °C, (*3*) 100 °C, (*4*) 120 °C. **(b)** Absorption spectra of $Cu(acac)_2$ in gaseous state (ca. 1.5×10^{-3} M, d = 2.8 cm). Temp.: 215 °C. Approximate λ_{max} values (nm) are given on the curves. (Taken from Mizutani et al. [12])

completely vacant. The spectrum they observed, shown in Fig. D.4b, can be taken as a strong deformation of the spectrum of the same compound in a nonpolar solvent (e.g. $CHCl_3$); the band is shifted to the red and is extremely broadened. The effects of thermal agitation, which weakens the effective ligand field, and the overlap of the band with the tail of the UV absorption, seem to be the reasons for these spectral changes. Since the temperature of the gas (215 °C) was quite high, a gradual decomposition took place, which lessened the accuracy of the data. However, the data on the gas-phase spectra of [$Cu(hfac)_2$], which is much more volatile, at 160–180 °C also support these views.

It may be added that the original aim of Mizutani et al. was to find clues to explain the difference between the colors of acetylacetonates in solid and gaseous states. Crystals of [Cu(acac)$_2$] and [Cr(acac)$_3$] are bluish violet and reddish violet, respectively, while their vapors are green. [Co(acac)$_2$] is pink, while its vapor is blue. The green color of [Cr(acac)$_3$] was found to be the result of a slight red shift in the absorption caused by a weakening of the ligand field at higher temperatures. The blue color of [Co(acac)$_2$] suggests that the gas is composed of tetrahedral monomers in contrast to the polymeric (tetrameric) octahedral structure of the pink solid [13], but thermal decomposition hinders a closer study. The spectra of these and related chelates in the gaseous state are, at any rate, an interesting field for further studies.

D.II.2 Tetrammine-type Complexes

The spectra of copper(II)-ammine and ethylenediamine complexes were first studied by Bjerrum and Nielsen [14] in connection with Bjerrum's classical study of the stepwise formation of complexes. Figure D.5a shows the spectra of $[Cu(NH_3)_n]^{2+}$ (n=1,... 5; coordinated water molecules are omitted in these formulae) obtained by analyzing the spectra of Cu^{2+}-NH_3 solutions.

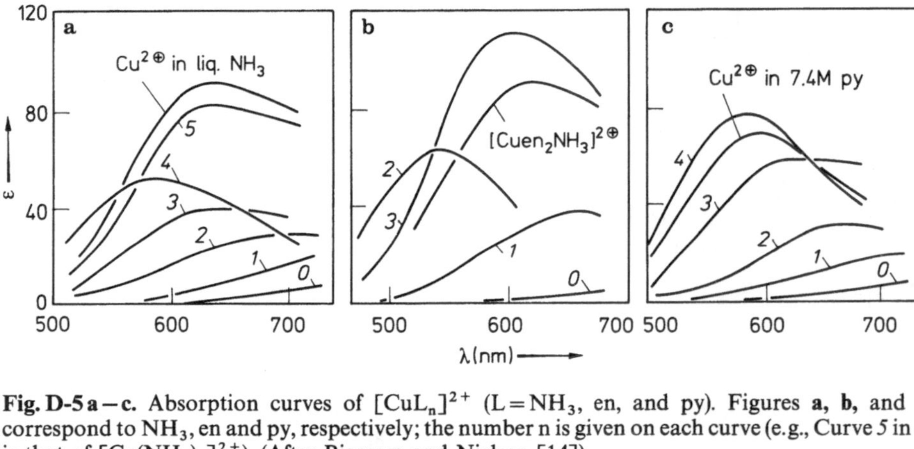

Fig. D-5a–c. Absorption curves of $[CuL_n]^{2+}$ (L=NH_3, en, and py). Figures **a**, **b**, and **c** correspond to NH_3, en and py, respectively; the number n is given on each curve (e.g., Curve *5* in **a** is that of $[Cu(NH_3)_5]^{2+}$). (After Bjerrum and Nielsen [14])

One notices that d-d band is shifted to the blue and increases in intensity as n increases from 1 to 4. The complex $[Cu(NH_3)_4]^{2+}$ is predominant in the usual bluish-violet solutions of Cu^{2+} and an excess of NH_3. Since NH_3 is a stronger ligand than H_2O, it prefers to substitute the water molecules occupying the equatorial sites of $[Cu(H_2O)_6]^{2+}$ first (the solution X-ray studies by Ohtaki and Maeda [15] have shown that $[Cu(H_2O)_6]^{2+}$ consists of an elongated octahedron, with four equatorial Cu-O bonds of 194 pm and two axial ones of 243 pm). The equatorial ligand field strength increases, step by step, in going from $[Cu(H_2O)_6]^{2+}$ to $[Cu(NH_3)_4(H_2O)_2]^{2+}$, thus shifting the d-d band to the blue.

With a large excess of NH_3, however, an additional NH_3 molecule is bound to the Cu^{2+} ion, forming $[Cu(NH_3)_5]^{2+}$. The d-d band is shifted back to the red and increases significantly in intensity. These spectral changes, called "pentaammine effect" by Jørgensen [16], are undoubtably due to the fact that the fifth NH_3 molecule must occupy one of the axial sites. The resulting increase in axial ligand field strength shifts the d-d band back to the red. Since the new axial ligand is the same as the equatorial ones, and both are quite strong, the d_{z^2} electrons will be squeezed out of the equatorial plane as shown in Fig. D.3. The coordination of water from behind the plane will be strongly weakened, and the complex adopts a nearly 5-coordinate structure (as assumed for [Cu(acac)$_2$] in pyridine). The coordination of a sixth molecule of NH_3 becomes almost impossible, so that even in liquid NH_3 cupric ions yield nearly the same spectrum as that of $[Cu(NH_3)_5]^{2+}$ in water.

Figure D.5b shows the curves for Cu^{2+}-en systems. In this case the equilibrium is much simpler. The addition of one and two moles of en to a solution containing one mole of Cu^{2+} leads to the almost quantitative formation of $[Cu(en)]^{2+}$ and $[Cu(en)_2]^{2+}$, respectively. The d-d bands are shifted to the blue. In a large excess of en, the octahedral complex $[Cu(en)_3]^{2+}$ is formed [17], again shifting the band to the red. Since en is a very strong chelating ligand, the three molecules must strongly compete for the equatorial positions. If a single molecule already occupies such a position, the other two must be strongly strained, due to an elongation of the axial bonds (remember that the d_{z^2} electrons repel the ligand atoms facing them!). Thus the most reasonable solution for all ligands is to make a compromise, i.e., not to distinguish between axial and equatorial positions in the octahedral coordination sphere. So the complex $[Cu(en)_3]^{2+}$ appears highly symmetrical in shape, but is in fact strongly "deformed", since the axial and equatorial bonds are compressed and elongated, respectively, in comparison to the length predicted by a Jahn-Teller distortion (Fig. D.1b). This is the reason why this chelate is much more unstable than $[Cu(en)]^{2+}$ or $[Cu(en)_2]^{2+}$ and its d-d band appears at the longer-wavelength side of $[Cu(en)_2]^{2+}$; the values of log K_1, log K_2 and log K_3 for the system Cu^{2+}-en are ca. 10.5, 9.3 and 1.0, respectively [18].

In addition to the above-mentioned trend which causes a static averaging of the Cu-N distances, a dynamic Jahn-Teller deformation will also result in a dynamic averaging of the same distances.

Figure D.5b also shows the curve for $[Cu(en)_2(NH_3)]^{2+}$, formed when $[Cu(en)_2]^{2+}$ is dissolved in concentrated NH_3. The influence of the pentaamine effect is again evident. A similar effect can also be observed by adding OH^- to $[Cu(en)_2]^{2+}$, whereby $[Cu(en)_2(OH)]^+$ is gradually formed [19].

All of these studies were carried out in aqueous media. Since the chelates $[Cu(en)_2]X_2$ are highly ionic, most of them are not sufficiently soluble in organic solvents (especially the aprotic ones). As a result chromotropic studies of these chelates are not as easy to carry out as in the case of [Cu(acac)$_2$]. However, Sone and Utsuno made some spectral observations, the results of which are summarized in Table D.1 [20].

Various kinds of information can be obtained from these data. It can be seen that the spectrum of $[Cu(en)_2]X_2$ does not depend on the nature of X^- in dilute aqueous solutions, since the anion are dissociated completely. The d-d band is observed at 550 nm ($\varepsilon = 64$). The band appears nearly at the same position in C_2H_5OH when $X^- = ClO_4^-$ or NO_3^-, and also in acetone when $X^- = ClO_4^-$. The axial influences of H_2O, C_2H_5OH or acetone are thus nearly the same. In CH_3NO_2, however, the d-d band is shifted to the blue, due to the weaker ligand field strength (or lower DN) of the solvent. It is interesting to note that the addition of large amounts of $NaClO_4$ or

Table D.1. Absorption maxima of $[Cu(en)_2]X_2$ in various solvents. (Taken from Sone and Utsuno [20]).

X	Solvent	λ_{max}(nm)	ε_{max}
ClO_4^-	H_2O	550	64
	C_2H_5OH	547	63
	acetone	547	63
	CH_3NO_2	534	60
	$H_2O + NaClO_4$(4M)	538	61
	$H_2O + NaClO_4$(8M)	528	61
	$H_2O + Mg(ClO_4)_2$(4M)	530	58
	py	582	96
NO_3^-	H_2O	550	64
	C_2H_5OH	547	63
	$H_2O + NaNO_3$(8M)	540	62
Cl^-	H_2O	550	64
	C_2H_5OH	575	84
	H_2O + LiCl(10M)	553	73

Temperature: 20–25 °C; chelate concentration: $7 \times 10^{-3} - 8 \times 10^{-3}$ M (7×10^{-4} M in the case of ethanol).

$Mg(ClO_4)_2$ (and to a smaller extent, $NaNO_3$) to an aqueous solution brings about similar spectral changes. This is evidently due to the "inert salt effect", i. e., a weakening of the axial Cu---OH_2 bonds caused by the solvation of the added ions.

On the other hand, a strong shift of the band in the opposite direction with a drastic increase in intensity occurs in pyridine, probably due to the formation of $[Cu(en)_2(py)]^{2+}$ (pentaamine effect). A similar shift is observed in a solution of $[Cu(en)_2]Cl_2$ in ethanol and, to some extent, in aqueous solutions containing large excesses of LiCl. These shifts are probably due to the formation of $[Cu(en)_2Cl]^+$. In the former case, the coordinating ability of Cl^- (which is poor in H_2O) is significantly increased in C_2H_5OH (which solvates Cl^- more weakly), while in the latter case some of the LiCl may act as an inert salt to dehydrate the cation and facilitate the coordination of Cl^- ions.

In all of these cases, the shape of the d-d band is reasonably symmetrical, i.e., any notable splitting is absent. Hence, the comparison of spectra measured under different conditions is apparently simple. In discussing these spectra it should be noted that, while the axial and equatorial ligand field strengths are different quantities, they are mutually correlated. For example, an incoming axial ligand of strong coordinating power weakens the bonds to the equatorial ligands somewhat, so that their ligand field strength is weakened. Or, when the ligand field strength of the equatorial ligands is weak, the positive charge on Cu^{2+} can not be effectively cancelled by their charges. Hence, the Cu^{2+} ion will attract the axial ligands more strongly and increase their ligand field strength. Moreover, the effectiveness of axial ligands is strongly influenced by the solvent. The less polar the solvent, the less is the Cu^{2+}-ligand interactin disturbed by it, and the stronger is the axial ligand field. It is conceivable that the equatorial ligands are also influenced by the polarity of the solvent; but this influence is presumably much less than that on the axial ones, since their coordinate bonds are stronger.

These situations can be more clearly examined in the case of the mixed chelates containing diamine and β-diketonato ligands to be discussed in the next section.

D.III Mixed Diamine-β-Diketonate Chelates

D.III.1 Introduction

Since both the bis-β-diketonato and bis-diamine chelates of Cu(II) show so many chromotropic color phenomena, it is to be expected that their mixed species, i.e., the chelates of the type [Cu(diam)(dike)]X, will also exhibit such phenomena. Fukuda, Sone et al. synthesized many such chelates and observed that they are soluble in a large number of solvents and exhibit marvelous solvatochromic phenomena [21–26, 27].

D.III.2 Chelate Perchlorates and Nitrates: Solute-Solvent Interactions

When N,N′-tetramethylethylenediamine and acetylacetonate ion (tmen and acac; for the abbreviations of other ligands to appear in the following sections, cf. Table C.3) are used as ligands, and ClO_4^- or NO_3^- as the anion X^-, the aqueous solutions of the resulting chelates [Cu(tmen)(acac)]X are bluish violet. The solutions become more and more reddish with decreasing solvent polarity (expressed in terms of DN). The DCE and CH_3NO_2(DN = 0 and 2.7, respectively) solutions are nearly red. Vice versa, the solutions become more and more bluish with increasing solvent polarity. The DMSO (DN = 29.8) solution is completely blue. In an interesting review on the indicators of solvent polarity, Soukup and Schmid [27a] pointed out that one can use the color of these chelates to estimate the value of the DN of the solvent, just as in the case of the chelate perchlorates and tetraphenylborates of Ni(II) (cf. Chap. C.IV.2).

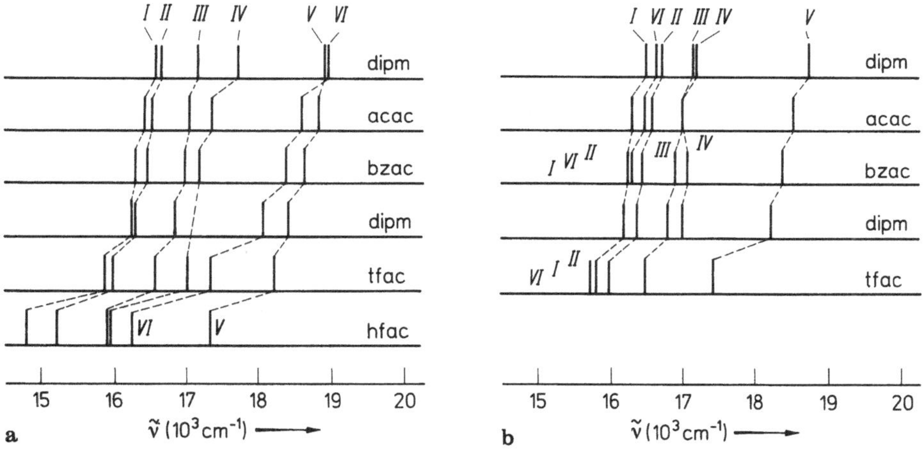

Fig. D-6 a, b. Comparison of the $\tilde{\nu}_{max}$ values of the chelates [Cu(tmen)(dike)]ClO_4 **(a)** and [Cu(tmem)(dike)]NO_3 **(b)** in various solvents. The symbols *I–VI* correspond, respectively, to DMSO (*I*; DN 29.8), DMF (*II*; 26.6), CH_3OH (*III*; 19), acetone (*IV*; 17.0), CH_3NO_2 (*V*; 2.7), and DCE (*VI*; 0). Conc.: ca. 5×10^{-3} M. (Taken from Sone and Fukuda [27])

A number of insights into the solute-solvent interactions in these solutions can be gained from the study of their absorption spectra. Figure D.6a shows the $\tilde{\nu}_{max}$ values of the d-d band of [Cu(tmen)(acac)]ClO_4 in various solvents, together with those for the chelates of the same type with different β-diketonato ligands. In most cases, there is a notable correlation between $\tilde{\nu}_{max}$ and DN, i.e., $\tilde{\nu}_{max}$ tends to increase in the order of decreasing DN.

Soukup and Schmid [27a] compared the absorption spectra of [Cu(tmen)(acac)]ClO_4 in 12 solvents. They found that there is a reasonably good linear relation,

$$DN = 195.5 - 0.0102\,(\tilde{\nu}_{max}),$$

between these quantities, confirming its excellent DN-indicating function.

It is also clear that the entire comb-like pattern for [Cu(tmen)(dipm)]ClO_4 shown at the top of Fig. D.6a is increasingly shifted to the red in the following order of β-diketonato ligand:

$$\text{dipm} \rightarrow \text{acac} \rightarrow \text{bzac} \rightarrow \text{dibm} \rightarrow \text{tfac} \rightarrow \text{hfac}. \qquad \text{(Eq. D.2)}$$

Only a few exceptions to these rules are observed. The most notable one involves the $\tilde{\nu}_{max}$ values in DCE. When dike is equal to dipm, the $\tilde{\nu}_{max}$ in this solvent is nearly the same as that in CH_3NO_2 (smmbol V). However, the former is shifted to the red much more drastically than the latter in the order given by (Eq. D.2). When dike is equal to hfac, the two values lie far apart. The $\tilde{\nu}_{max}$ values observed with [Cu(tmen)(dike)]NO_3 shown in Fig. D.6b are similar to those given in Fig. D.6a; here the anomaly in DCE is much more evident.

All of these phenomena can be explained reasonably in the following way. These complexes are strongly dissociated, almost always into the chelate cations $[Cu(tmen)(dike)]^+$ and ClO_4^- or NO_3^-. The chelate cation formed is axially solvated, so that its shape is transformed from the original square planer one to an elongated octahedral, and eventually to a pseudo octahedral one, depending on the strength of solvation or donor ability of the solvent molecule. Figure D.7 illustrates these situations schematically.

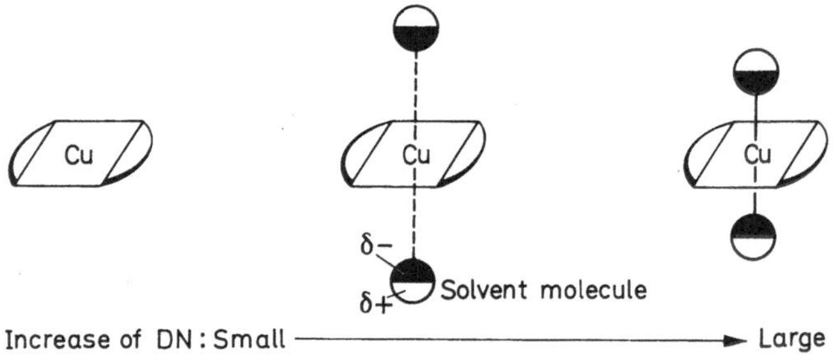

Fig. D-7. Axial solvation of the chelate cation, $[Cu(diam)(dike)]^+$. (Taken from Sone and Fukuda [27])

The situations are therefore similar to those of $[Cu(acac)_2]$ described in Chap. D.II. So the apparent $\tilde{\nu}_{max}$ of the chelate decreases with the increase of the DN of the solvent, i.e., with that of axial solvation. The correlation of the $\tilde{\nu}_{max}$ values with the nature of the dike (Eq. D.2) can also be explained in terms of the same model. In fact, this order is nearly the same as the one found for the ease of formation of the octahedral form of Ni(II) chelates (cf. Chap. C.IV.2), which can be attributed to the electron-releasing or electron-withdrawing effects of the substituents on dike (Hammett constants). Thus, the electron releasing effect of the two tert-butyl groups in the dipm chelate makes the chelate ring more negative. The donor strength (or ligand field strength) of dike increases, and the coordination sphere about the Cu(II) ion becomes more electron-rich. The approach of solvent molecules in the axial directions gets more difficult than with the acac chelate. All these effects will therefore shift the $\tilde{\nu}_{max}$ values towards the blue, i.e., towards a less octahedral, more square planar structure. Just the opposite can be said for the tfac or hfac chelates with their electron-withdrawing groups. As expected, the phenyl-substituted bzac and dibm ligands exhibit an intermediate behaviour between dipm/acac and tfac/hfac, favoring axial coordination slightly more than acac (cf. the Hammett constants of CH_3 and C_6H_5; Chap. C.IV.2). Figure D.8 is a schematic representation of these effects.

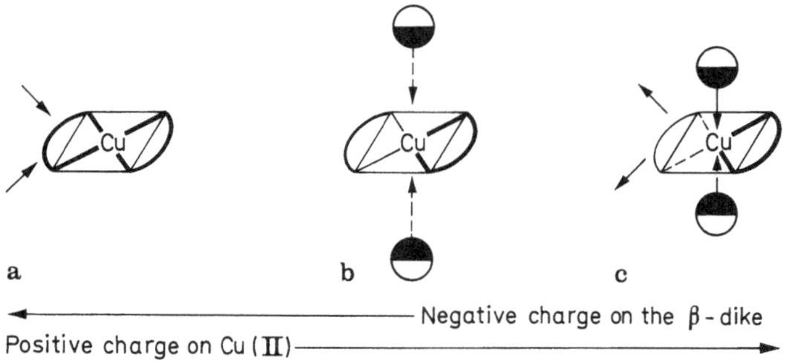

Fig. D-8 a–c. Schematic representation of the substituent effects on dike. **(a), (b)** and **(c)** are three chelates, with electron-releasing, intermediate and electron-withdrawing substituents. (Taken from Sone and Fukuda [27])

The anomaly in DCE is seemingly the result of a weak coordination of ClO_4^- or NO_3^-, which takes place in this solvent of low dielectric constant ($\varepsilon/\varepsilon_0$: 10.1; note that this value is above 20 in most of the other solvents). If this takes place at an axial coordination site to yield a 5-coordinate structure, the $\tilde{\nu}_{max}$ of the complex will be shifted to the red, since the spectral effect will be essentially similar to that of axial solvation. On the other hand, such a weak coordination is less likely in CH_3NO_2, whose DN (2.7) is nearly as low as that of DCE but whose dielectric constant (36.7) is much higher. So the chelate cation remains nearly square planar. The fact that the difference between the $\tilde{\nu}_{max}$ in DCE and that in CH_3NO_2 increases with decreasing ligand field strength of dike, and that the same difference increases notably when ClO_4^- is substituted by NO_3^- which is of higher coordination ability, is consistent with this explanation. Conductivity data on some of these solutions also support this idea.

It is also of interest to note the influence of the substituents R_1-R_4 (Table C.3) on the diamine ligand. Figure D.9 shows some of the relevant data.

As these substituents become more and more bulky, the range between I and V–VI expands; that is to say, the square planar species becomes more strictly planar, while the octahedral species becomes more regularly octahedral. In the case of teen, in which

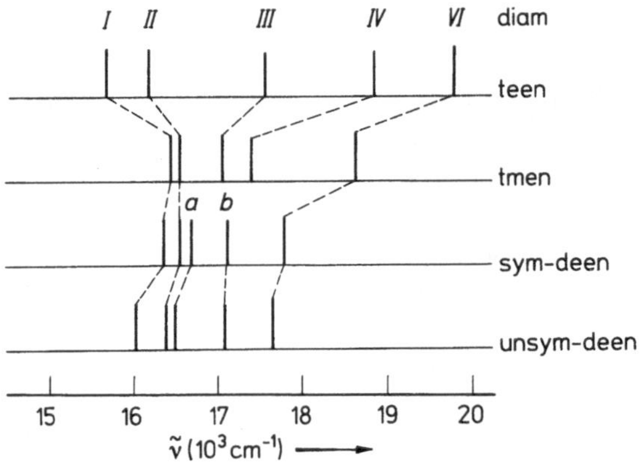

Fig. D-9. Comparison of the $\tilde{\nu}_{max}$ values of four chelates [Cu(diam)(acac)]ClO_4. The symbols *I* to *VI* used in this figure are the same as those in Fig. D-6; a: H_2O; b: CH_3CN. (Taken from Sone and Fukuda [27])

all of the substituents are bulky ethyl groups, the coordination of solvents of low DN at the axial sites is strongly hindered. The resulting species thus becomes very planar. On the other hand, solvents of higher DN tend to get coordinated by any means, getting through the ethyl groups protruding above and below the plane of the chelate. Hence, the equatorial ligand field strength will be weakened by interligand repulsion, making the species highly octahedral. With deen's the axial sites are more open; the differences among the effects of the solvents diminish and a narrow range from I to VI is observed.

In this way, we can study a number of interesting solute-solvent interactions with these mixed-chelate cations, which can generally be systematized using the concept of DN.

D.III.3 Halogeno- and Pseudohalogeno-Complexes: Solute-Solute-Solvent Interactions

When the anion X^- in the complex Cu(tmen)(acac)X is a halide (Cl^-, Br^- or I^-) or a pseudohalide (NCS^-, N_3^- or NCO^-), a new type of spectral change due to solute-solute-solvent interactions comes into play. These anions are much better ligands than ClO_4^- or NO_3^-, so that they now compete with the organic solvents for the axial sites of the chelate cation.

Again the complexes are soluble in many organic solvents, but the observed colors of the solutions are usually quite different from those observed with the perchlorate and nitrate of the same chelate cation. Even the colors of the solids are markedly different; the halogeno- and pseudohalogeno-complexes are in various shades of green, while the perchlorate and nitrate are bluish violet. The colors of the aqueous and methanolic solutions of all these complexes are bluish violet, indicating that they are all strongly ionized in such solvents into the axially solvated chelate cation $[Cu(tmen)(acac)]^+$ and the corresponding anion. On the other hand, the solutions of the complex halides or pseudohalides in other solvents generally range from a bluish color to a greenish in the following order:

H_2O/MeOH → EtOH → DMSO → DMF → CH_3NO_2/CH_3CN → acetone/DCE
bluish violet ⟶ blue ⟶ greenish blue ⟶ green (Eq. D.3)

It is interesting to note that this order is very different from that observed for DN. It is much better correlated with certain other polarity parameters of these solvents, such as Gutmann-Mayer's acceptor number (AN) [28], Kosower's Z [29], or Dimroth-Reichardt's E_T [30].

Some comments will now be given on the nature of these three scales, for a better understanding of the underlying situations.

Of these three parameters, the meaning of AN is probably the simplest to understand. This parameter is a scale for the magnitude of ^{31}P-NMR chemical shift of Et_3PO in a certain solvent. It is a measure of the ability of a solvent to solvate a negative particle (an anion or the negative end of a polar solute molecule). The P atom of Et_3PO is effectively shielded from external influences by its three bulky ethyl groups. It is only susceptible to the effect of the solvent molecules solvating the oxygen atom peeping out from the ethyl groups. Thus, it is natural to assume that the chemical shift of the P atom, which reflects the electronic distribution about it, changes with the strength of this solvation. Here, the solvent molecules serve as acceptors for the electron pairs on the oxygen atom. So it is called acceptor number.

The two parameters, Z and E_T, are very similar to each other in their general ideas. A linear relation exists between their values ($Z = 1.259\,E_T + 13.76$ [31]). They were determined from the spectral studies of two highly polar and solvatochromic organic compounds, I and II, shown in Fig. D.10.

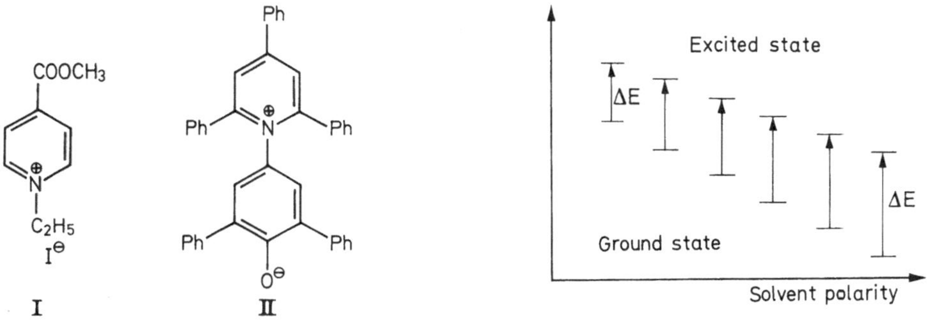

Fig. D-10. Organic dyes used in the determination of Z(I) and E_T(II), and the relation between ΔE (=hν) and solvent polarity

Each of these compounds shows a strong absorption maximum in the visible region, which shifts remarkably towards the blue (or ultraviolet) with an increase in solvent polarity. The energy of this absorption, measured in a certain solvent and expressed in kcal mol^{-1}, is taken as the Z (with I) or E_T (with II) value of that solvent, respectively.

It is believed that the electronic excitation leading to this absorption involves a remarkable charge transfer (CT) in the molecule. In the case of I, an electron jumps from the I^- ion to the positively charged pyridine ring,

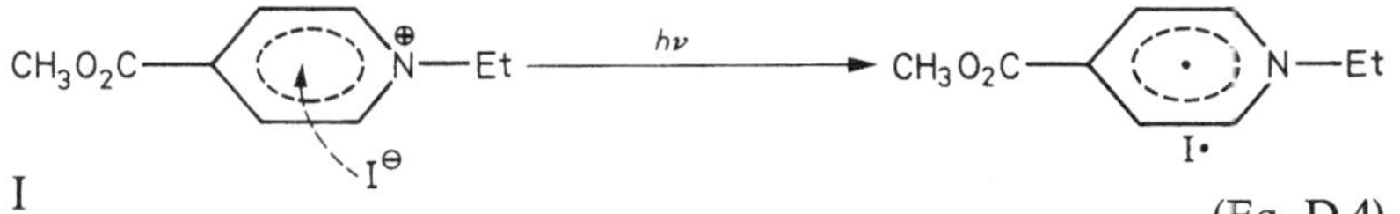

(Eq. D.4)

and is captured by the π-electron orbital. In II, the charge on the oxygen atom moves toward the highly conjugated π-electron system:

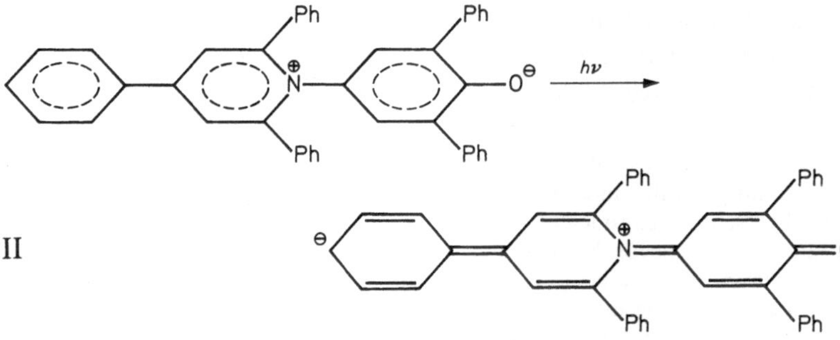

II

(Eq. D.5)

Only one of the possible resonance structures of the excited state is shown in (Eq. D.5). A large number of such structures, with the negative charge localized on the other carbon atoms of the seven aromatic rings, can easily be written down. One can see that the negative charge, concentrated on the oxygen atom in the ground state, is smeared out over the entire molecule in the excited state. At any rate, the ground states of these molecules are much more polar than their excited states, so that the former are more strongly stabilized by the solvation than the latter, as shown diagrammatically in Fig. D.10. This is the reason why the excitation energies of these systems, ΔE, can be taken as a measure of the solvent polarity.

Using these polarity scales, we can arrange common polar solvents in the following order of decreasing polarity:

Solvent	H_2O →	MeOH →	EtOH →	CH_3NO_2 →	CH_3CN →	DMSO →
AN	54.8	41.3	37.1	20.5	19.3	19.3
	DCE →	DMF →	Acetone			(Eq. D.6)
	16.7	16.0	12.5			

Solvent	H_2O →	MeOH →	EtOH →	CH_3NO_2 →	CH_3CN →	
$E_T(Z)$	63.1(94.6)	55.5(83.6)	51.9(79.6)	46.3(−−)	46.0(71.3)	
	DMSO →	DMF →	Acetone →	DCE		
	45.0(71.1)	43.8(68.5)	42.2(65.7)	41.9(−−)		(Eq. D.7)

It can be seen that these two orders are nearly the same, with the exception of DCE, which appears relatively high in the AN scale but very low in the E_T or Z scale. Such a general agreement is what we can expect. In the ground-state formulae of I and II, the negative charges are concentrated on the iodine or oxygen atom, while the positive charges are not localized on the nitrogen atom, but are more or less smeared over the surrounding π-electron systems. The solvation of such systems therefore mainly takes place at the negatively charged atom of the solute. The strength of such an interaction is naturally governed by the AN.

We can now return to the problem of the mixed chelates, Cu(tmen)(acac)X (X^- = halide or pseudohalide anion). We stated that the color of the solutions of these complexes changes from bluish violet to green in the order of the solvents given in (Eq. D.3), which is reasonably similar to the decreasing order of AN, E_T or Z of the solvents. Spectral studies of these solutions, the results of which are summarized in Fig. D.11, also reveal the same correlations.

One can see that the $\tilde{\nu}_{max}$ values in H_2O and MeOH do not depend on the anion X^-. Except for the iodide, the absorption maxima in other solvents shift towards the

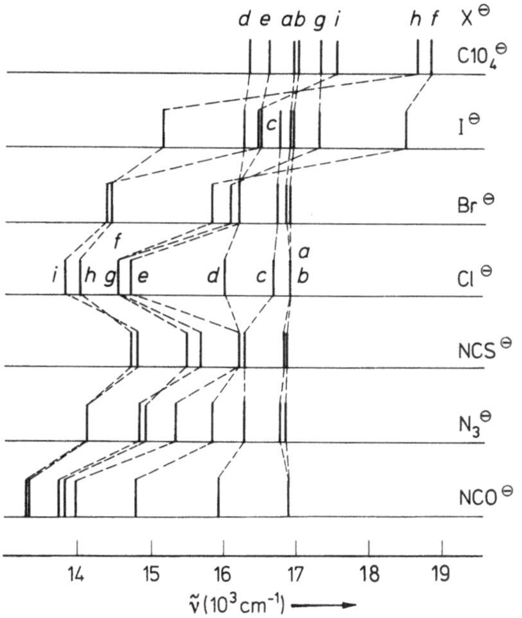

Fig. D-11. Comparison of the $\tilde{\nu}_{max}$ values of the chelates [Cu(tmen)(acac)]X (X^- = halide or pseudohalide ion). The data for the chelate perchlorate are also given for a comparison. The symbols *a* to *i* correspond to H_2O (*a*), CH_3OH (*b*), C_2H_5OH (*c*), DMSO (*d*), DMF (*e*), CH_3NO_2 (*f*), CH_3CN (*g*), DCE (*h*), and acetone (*i*), respectively. Conc.: ca. 5×10^{-3} M. (Taken from Sone and Fukuda [27])

red, generally in the order of (Eq. D.3). The shift tends to stop in DCE and acetone. It was also determined that the halogeno- and pseudohalogeno-complexes are practically non-electrolytes in the latter two solvents, while in H_2O and MeOH they are 1:1 electrolytes and all other solutions show intermediate conductivities. Careful analysis of these and related data leads to the following picture.

1) In DCE and acetone, these chelates exist as non-ionic, 5-coordinate complexes, i.e., [Cu(tmen)(acac)X].

2) In other solvents, these 5-coordinate complexes dissociate into solvated $[Cu(tmen)(acac)]^+$ and X^- as shown below. The degree of dissociation increases in the order opposite to that given in (Eq. D.3). It is nearly complete in H_2O an MeOH.

$$[Cu(tmen)(acac)X] \rightleftharpoons [Cu(tmen)(acac)]^+ + X^- \qquad \text{(Eq. D.8)}$$

	5-coordinate	tetragonal, solvated
	(green)	(blue; cf. Chap. D.III.2)
Solvent:	DCE/Acetone ----------	MeOH/H_2O
Polarity:	Small ⟵⟶	large
(AN, E_T or Z)		

The assumption that the neutral species formed in DCE or acetone are 5-coordinate complexes is supported by the similarity of the spectra of each compound in these two solvents. It was further observed that the $\tilde{\nu}_{max}$ of the solution of a chelate in a solvent of intermediate polarity, e.g., CH_3CN or CH_3NO_2, is gradually shifted to the red and approaches the value observed in DCE or acetone, when a salt MX (KNCS, $(C_2H_5)_4NCl$, etc.) containing the same anion as the chelate is added to the solution. This corresponds to a shift in the equilibrium of (Eq. D.8) to the left-hand side, and indicates again that the neutral species is not effectively solvated in any of these solvents, i.e., it is to be formulated as 5-coordinate [Cu(tmen)((acac)X]. The axial ligand field, produced by a single X^- ion, is seemingly much stronger than the one produced by solvent molecules; as a result, the $\tilde{\nu}_{max}$ values of [Cu(tmen)(acac)X] species are much lower than those of the solvated $[Cu(tmen)(acac)]^+$ ions observed in solutions of $[Cu(tmen)(acac)]ClO_4$. Moreover, the band of the former is much stronger than that of the latter, as is often observed in the formation of 5-coordinate complexes in solution (cf. Fig. D.5).

A comparison of the spectral profiles supports further the view that the absorption band of [Cu(tmen)(acac)X] in a solvent of intermediate polarity is made up of the d-d bands of the two species coexisting in the above equilibrium. Its apparent $\tilde{\nu}_{max}$ decreases with a shift in the equilibrium to the left-hand side as stated above. Analyses of the spectral changes caused by the addition of a salt MX, or by the changes in the concentration of the complex, as well as the analysis of the conductivity data, are consistent with this view; in some cases, the values of K for the equilibrium can be estimated with reasonable accuracy.

Now, Fig. D.11 indicates that the coordination ability of the anions at the axial site of $[Cu(tmen)(acac)]^+$ increases in the following order:

$$I^- < Br^- \cong NCS^- < Cl^- \cong N_3^- < NCO^-. \qquad \text{(Eq. D.9)}$$

The I^- ion is particularly reluctant to be combined in this way. The spectra of the iodide complex are apparently irregular; in a number of solvents the iodide ion behaves somewhat like the perchlorate ion. Only when the polarity of the solvent is very low (e.g., in DCE) does it behave like the other halides.

The orders, $I^- < Br^- < Cl^-$ and $NCS^- < N_3^- < NCO^-$, considered separately, are quite conceivable and comparable with the results obtained with other complexes [32]. However, the entire order given in (Eq. D.9) may seem somewhat strange, because usually pseudohalide ions are considered to be stronger ligands than the halide ions. This may be the result of a poor solvation of the spherical, inert-gas-like halide ions in organic solvents, which makes them much stronger ligands in such solvents than in water (in the case of the pseudohalide ions, their linear, π-conjugate structure will make them more solvated and weaker ligands than the halide ions in organic solvents). Although only little information is available on this problem, the data of Ahrland on relevant complex stabilities and solvation energies [33–34, 34a] seem to be compatible with this view.

When such a 5-coordinate chelate molecule is attacked by polar solvent molecules, the resulting reaction (i.e., the forward reaction in (Eq. D.8) will be somewhat like that shown in Fig. D.12. Some solvent molecules will surround the X^- ion coordinated to the chelate plane, probably on the top of a square pyramidal structure, and pull out the anion from the coordination sphere. Other solvent molecules will try to attack the Cu(II) ion directly to form a solvated chelate cation.

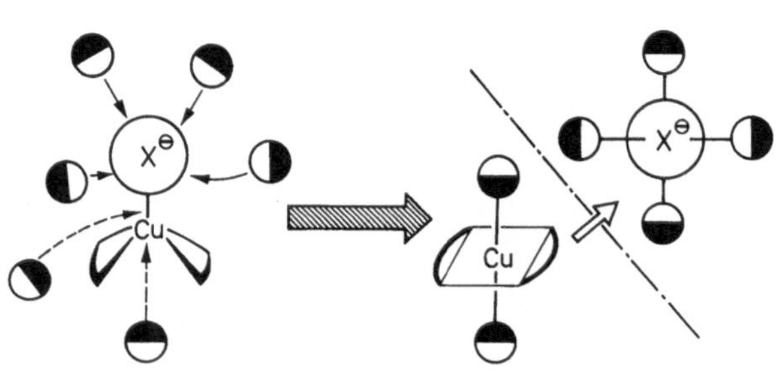

Fig. D-12. Proposed mechanism for the reaction [Cu(tmen)(acac)X] → $[Cu(tmen)(acac)]^+$ (Solvated) $+X^-$ in a polar organic solvent. (Taken from Fukuda et al. [36])

This situation is apparently very different from the one depicted in Fig. D.7, in which the naked solute, i.e., the chelate cation, is open to a direct attack by the solvent molecules. The solvent molecules must, first of all, drive out the firmly bound X^- anion from the coordination sphere. After this step has been passed, the chelate cation will be readily solvated. Thus, the most important characteristic of the solvent for a color change (or shift in equilibrium) must be its power to solvate the anion, i.e., its acceptor ability expressed in terms of AN or related parameters (E_T/Z).

The observed order of spectral shifts given in (Eq. D.3) is thus similar to those of AN, E_T and Z. The readers may have already noticed that these orders are not exactly the same. For example, DMSO and DMF, which should appear after CH_3NO_2 and CH_3CN according to these three parameters, appear before the latter solvents in (Eq. D.3). Also DCE, which should appear near to DMF according to the AN scale, appears at the end with acetone in (Eq. D.3), as expected from the E_T scale. These discrepancies can now be accounted for, by taking the two roles of the solvent molecules shown in Fig. D.12 into account. As acceptors the solvent molecules pull out the anion X^-, whereas as donors they solvate the complex cation. In other words, we can expect the reaction shown in Fig. D.12 to be governed, not only by AN, but by an appropriate combination of AN and DN, such as the polarity factor (Pf) given below [35, 36]:

$$Pf = a(AN) + b(DN) \qquad \text{(Eq. D.10)}$$

In the present case, the contribution of the first term, a(AN), in this new parameter is notably larger than that of the second term, b(DN); the general order of Pf is therefore similar to that of AN. However, when DN is very small (0 in the case of DCE) or very large (29.8 and 26.6 with DMSO or DMF, respectively), the value of Pf accordingly becomes smaller or larger, leading to their respective positions in the observed order (Eq. D.3).

It is possible that a similar effect of DN is also reflected in the position of DCE on the E_T scale. As pointed out above, E_T is mainly a measure of the solvation of the negative group of an organic dye (II, Fig. D.10). It includes, however, the contribution from the solvation of its positive group to some extent, so that a solvent with a very small DN, such as DCE, may exhibit an E_T which is much smaller than that expected from its AN.

The fact that the above-mentioned equilibrium is really governed by the factor Pf is illustrated in Fig. D.13, where the $\tilde{\nu}_{max}$ values of the solutions of [Cu(tmen)(tfac)X]

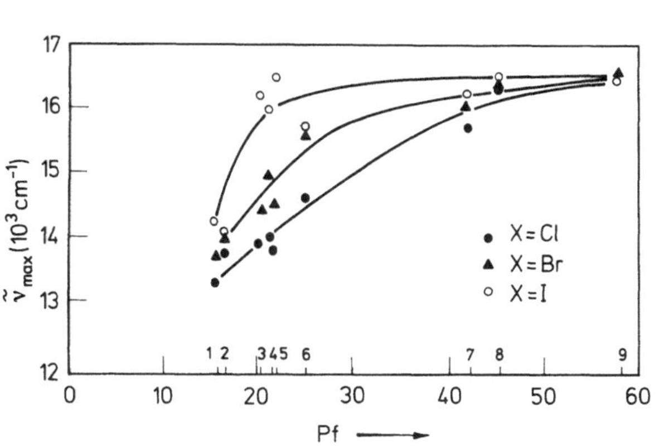

Fig. D-13. The correlation between the $\tilde{\nu}_{max}$ values of Cu(tmen)(tfac)X in various solvents and the solvent parameter Pf. The points *1* to *9* on the Pf scale correspond to acetone, DCE, CH_3NO_2, DMF, CH_3CN, DMSO, EtOH, MeOH, and H_2O, respectively. (After Fukuda et al. [36])

($X^-=Cl^-$, Br^-, I^-) in various solvents are plotted against tentative values of Pf (Pf = 1.00(AN) + 0.20(DN)) [36]. It can be seen that the value of $\tilde{\nu}_{max}$ increases gradually with an increase of Pf if $X^- = Cl^-$, indicating that the equilibrium is shifted in the same way towards the right-hand side. With I^-, on the other hand, $\tilde{\nu}_{max}$ rises steeply while Pf is still relatively small and quickly reaches a plateau, the height of which is nearly the same as that of the right-hand end of the curve for Cl^-. This shows that I^- ions are only weakly coordinated to Cu(II), so that they are easily driven out of the coordination sphere by a small increase in Pf. The anion Br^- lies between these two extremes.

It is of interest to study these equilibria in more detail, and also to study the thermochromic and piezochromic behaviors of these strongly solvatochromic systems. Preliminary studies in these directions have already yielded some interesting results. We hope that the study of these complexes and the corresponding nickel(II) chelates described in Chapter C will serve as small but attractive milestones for the travelers wandering in the land of inorganic chromotropism.

References

1. Huheey, J. E.: "Inorganic Chemistry", 3rd Ed., Harper, Cambridge (1983)
2. Hathaway, B. J., Tomlinson, A. A. G.: Coord. Chem. Rev. **5**, 1 (1970)
3. Hathaway, B. J., Billing, D. E.: Coord. Chem. Rev. **5**, 143 (1979)
4. Hathaway, B. J.: "Structure and Bonding", Vol. 14, Springer, Heidelberg (1973), p. 49
5. Gažo, J. et al.: Coord. Chem. Rev. **19**, 253 (1976)
6. Belford, R. L. et al.; J. Inorg. Nucl. Chem. **2**, 11 (1956)
7. Belford, R. L. et al.: J. Chem. Phys. **26**, 1165 (1957)
8. Koyama, H. et al.: J. Inst. Polytech., Osaka City Univ., 4, No. 1, Ser. C, 43 (1953)
9. Shibata, S., Sone, K.: Bull. Chem. Soc. Jpn. **29**, 852 (1956)
10. Graddon, D. P.: Nature **183**, 1610 (1959)
11. Yokoi, H. et al.: Bull. Chem. Soc. Jpn. **45**, 1100 (1972)
12. Mizutani, K. et al.: Z. anorg. allg. Chem. **365**, 217 (1969)
13. Cotton, F. A., Elder, R. C.: Inorg. Chem. **4**, 1145 (1965)
14. Bjerrum, J., Nielsen, E. J.: Acta Chem. Scand. **2**, 297 (1948)
15. Ohtaki, H., Maeda, M.: Bull. Chem. Soc. Jpn. **47**, 2197 (1974)
16. Jørgensen, C. K.: "Absorption Spectra and Chemical Bonding in Complexes", Pergamon, Oxford (1962), p. 125
17. Cullen, D. L., Lingafelter, E. C.: Inorg. Chem. **9**, 1858 (1970)
18. Sillén, L. G., Martell, A. E.: "Stability Constants of Metal-ion Complexes", The Chemical Society, London (1964), p. 119
19. Jonassen, H. B. et al.: J. Am. Chem. Soc. **77**, 2748 (1955)
20. Sone, K., Utsuno, S.: Bull. Chem. Soc. Jpn. **39**, 1813 (1966)
21. Fukuda, Y., Sone, K.: Bull. Chem. Soc. Jpn. **45**, 465 (1972)
22. Fukuda, Y. et al.: J. Inorg. Nucl. Chem. **36**, 1265 (1972)
23. Fukuda, Y. et al.: Bull. Chem. Soc. Jpn. **50**, 142 (1977)
24. Fukuda, Y. et al.: ibid. **50**, 313 (1977)
25. Fukuda, Y. et al.: ibid. **54**, 428 (1981); cf. footnote of ref. (26)
26. Fukuda, Y. et al.: ibid. **55**, 3738 (1982)
27. Sone, K., Fukuda, Y.: "Ions and Molecules in Solution", (Ed. Tanaka, N. et al.), Elsevier, Amsterdam (1983), p. 251

27a. Soukup, R. W., Schmid, R.: J. Chem. Educ. **62**, 459 (1985)

28. Gutmann, V.: "The Donor-Acceptor Approach to Molecular Interactions", Plenum, New York (1978)

29. Kosower, E. M.: An Introduction to Physical Organic Chemistry", Wiley, New York (1968)
30. Dimroth, K. et al.: Ann. Chem. **661**, 1 (1963)
31. Reichardt, C.: Angew. Chem. Internat. Ed. Engl. **4**, 29 (1965)
32. cf. e.g. Gutmann, V.: "Coordination Chemistry in Non-Aqueous Solutions", Springer, Wien (1968); note, however, that the order among the pseudohalide ions is still somewhat ambiguous
33. Ahrland, S.: "The Chemistry of Nonaqueous Solvents", Ed. J. J. Lagowsky, Vol. V A, Academic Press, New York (1978), p. 1
34. Ahrland, S.: Pure & Appl. Chem. **54**, 1451 (1982)
34a. Ahrland, S., Björk, N.O.: Coord. Chem. Rev. **16**, 113 (1975)
35. The usefulness of such a combination parameter in the study of intermolecular interactions in organic liquids was already demonstrated by Schmid; cf. Gutmann, V.: Coord. Chem. Rev. **43**, 150 (1982)
36. Fukuda, Y. et al.: Bull. Chem. Soc. Jpn. **58**, 3518 (1985)

CHAPTER E

Miscellaneous Chromotropic Phenomena Observed in Solutions of Metallic Complexes

E.I Introduction

A large number of chromotropic phenomena in solutions have also been observed and reported for transition metal complexes not covered in the foregoing chapters. In this Chapter we shall concentrate ourselves on several representative examples in which systematic studies led to a reasonably clear-cut understanding of the phenomena. It is interesting to note that the concepts of DN and AN, or of related polarity parameters, appear again and again as unifying principles in many of these examples; it is a proof of their general validity in explaining the changes that occur in solution.

E.II Spin-Crossover in Iron(II) and Iron(III) Complexes Observed as Solution Thermochromism

Many examples of thermochromic equilibria in which high-spin, octahedral chelates of Ni(II) are converted into low-spin, square-planar chelates were given in Chapter C. In these cases, both the spin state and the geometry of the chelates are changed simultaneously upon haeting or cooling.

This is in strong contrast to the Co(II) complexes or Cu(II) chelates, treated in Chap. B and D, respectively, where the changes in color are accompanied by the changes in geometry, but not by those in the spin state. Most Co(II) complexes are high-spin, whereas there is no distinction between high- and low-spin states for Cu(II).

On the other hand, thermochromic color changes, which are due to changes in the spin state, but not to any remarkable change in geometry, are sometimes observed with Fe(II) and Fe(III) complexes. Here both high- and low-spin complexes are common; $[Fe(H_2O)_6]^{2+}$ and $[Fe(H_2O)_6]^{3+}$ are high-spin, while $[Fe(CN)_6]^{4-}$ and $[Fe(CN)_6]^{3-}$ are low-spin. Among the chelates, for example, $[Fe(acac)_2]$ and $[Fe(acac)_3]$ are high-spin, while $[Fe(phen)_3]^{2+}$ and $[Fe(phen)_3]^{3+}$ (phen = 1,10-phenanthroline) are low-spin. It should be noted that most of these complexes are octahedral. The change from high- to low-spin is brought about by an increase in ligand field strength, as shown in Fig. E.1

If the ligand field strengths in these complexes are very carefully "tuned", by using specially designed ligands, it is sometimes possible to prepare complexes which are neither high-spin nor low-spin. That is to say, when the difference in energy between the high- and low-spin states is of the order of kT, the Fe^{2+} or Fe^{3+} ions in the complex

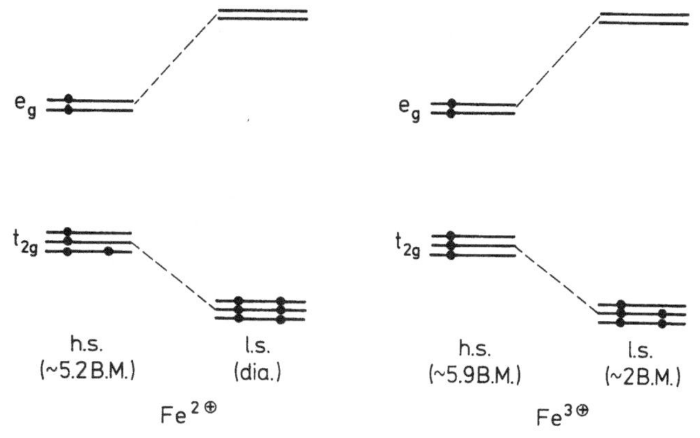

Fig. E-1. High-spin (h.s.) and low-spin (l.s.) states of Fe^{2+} and Fe^{3+} in octahedral ligand fields

will be partially high-spin and partially low-spin, and a thermal equilibrium is established among them.

Strongly temperature-dependent changes in the magnetic moment and the spectra are observed in the course of such a "spin crossover".

Although a number of such phenomena are known [1], most of them have been studied by means of magnetic measurements in the solid state, and have only rarely been observed as solution thermochromism. Recently, however, Chum et al. studied such a case with $[Fe(amp)_3]^{2+}$, where amp = 2-(aminomethyl)-pyridine [2]. They found that this chelate exhibits two intense absorption bands at ca. 440 nm and 380 nm in water or water/acetonitrile. The relative intensities of these bands change sharply with temperature (cf. Fig. E.2).

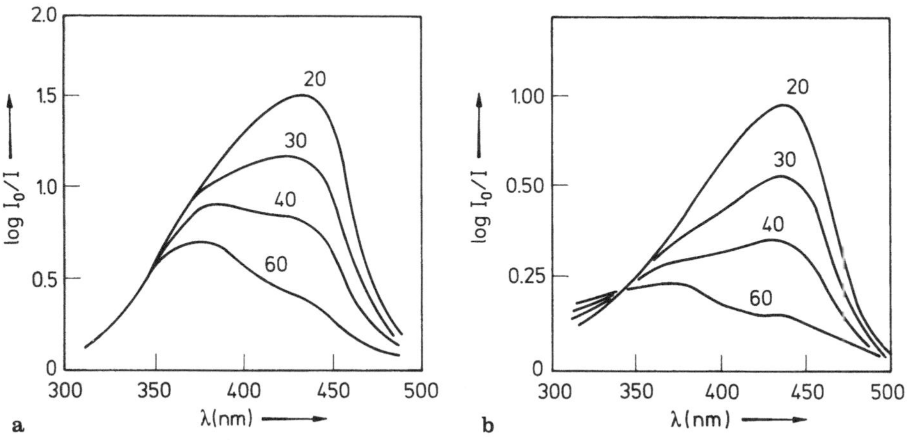

Fig. E-2 a, b. Absorption spectra of $[Fe(amp)_3]^{2+}$ at different temperatures. **(a)** 4×10^{-4} M in CH_3CN/H_2O/amp; **(b)** 1.9×10^{-4} M in H_2O/amp (pH 9). Temperature (°C) is given on each curve. (After Chum et al. [2])

Magnetic and ESR studies support the view that this thermochromism is associated with the spin crossover in the complex, which can be formulated as follows:

$$[Fe(amp)_3]^{2+}(t_{2g}{}^6, {}^1A_{1g}) \rightarrow [Fe(amp)_3]^{2+}(t_{2g}{}^4e_g{}^2, {}^5A_1) \quad \text{(Eq. E.1)}$$

The ΔH and ΔS values for this equilibrium in acetonitrile/H_2O were estimated to be 21.3 kJ mol^{-1} and 71 $JK^{-1} mol^{-1}$, respectively, from the spectral data. The molar extinction coefficients ε of these strong absorption bands lie in the range of several thousands, i.e., these absorption bands are much stronger than common d-d bands. They are M $\rightarrow$ L (metal to ligand) charge transfer bands which are frequently observed in the spectra of Fe(II) chelates (cf. Chap. E.II).

Similar observations have been made by Matsumoto et al. [3] with an Fe(III) chelate containing a planar and unsymmetrical tetradentate Schiff base (L) and imidazoles (Im).

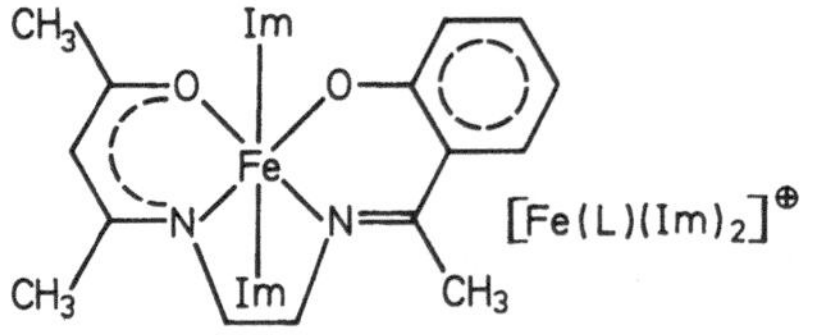

This chelate (obtained as $[Fe(L)(Im)_2]B(Ph)_4$) exhibits a brilliant reversible thermochromism, changing from dark red at room temperature to green at ca. $-70\,°C$ in various organic solvents. The dark red solutions exhibit a band at 520 nm (ε: 2500), while the green ones show a band of similar intensity at 685 nm. ESR studies have revealed that the former contain a high-spin chelate, while the latter contain a low-spin one. It is also interesting that the solid chelate itself is green and essentially low spin (μ_{eff}: 3.20 BM), exhibiting a reflection spectrum which is similar to the absorption spectra of the cooled green solutions. This is, therefore, another example of solution thermochromism caused by a spin crossover. The responsible bands are of the M $\leftarrow$ L (ligand to metal) type, commonly found among the spectra of Fe(III) chelates.

$$\underset{\text{green}}{[Fe(L)(Im)_2]^{+}(t_{2g}{}^5, S=1/2)} \rightarrow \underset{\text{red}}{[Fe(L)(Im)_2]^{+}(t_{2g}{}^3e_g{}^2, S=5/2)} \quad \text{(Eq. E.2)}$$

These examples of "crossover thermochromism", and several others cited in the above papers, are still small in number, but more examples are bound to be discovered [3a]. Spin crossover is also known to occur in certain complexes of Co(II) and Co(III).

E.III Iron(II) Cyanide-(Heterocyclic N-Base) Mixed Complexes: Solvatochromism Caused by Acceptor Solvents and Ions

In the last section, mention was made of the fact that a strong M $\rightarrow$ L type of CT band often appears in the spectra of Fe(II) complexes. Typical examples are the Fe(II) chelates of bipy (bipy = 2,2′-bipyridine) and phen. The stability and ease of formation of these deep red chelates in aqueous solutions are well known since their discovery by Blau in late 19th century. They have been widely used in the detection and colorimetric analysis of Fe(II). The origin of such a deep color was first discussed by Sone [4], who ascribed it to the existence of characteristic chromophoric groups, i.e., to the unsaturated rings (**A**; see below) in such chelates.

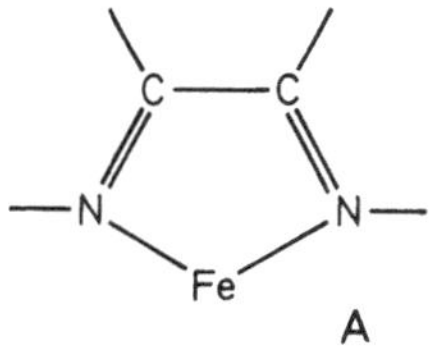

A

This view was soon confirmed by Krumholz [5], who synthesized a number of $[Fe(L)_3]^{2+}$-type chelates of aliphatic α-diimines, which contain three A-type chelate rings with four alkyl groups. They are deep violet and show absorption spectra very similar to those of $[Fe(bipy)_3]^{2+}$ and $[Fe(phen)_3]^{2+}$, just as expected from Sone's concept. Busch and Bailar [6], who also studied similar chelates, called the chromophoric group an "iron-methine chromophore". This concept was further developed by a large number of investigators. They eventually came to the conclusion that the strong absorption band of these chelates, which appears at 500–600 nm, is due to a CT process in which a fraction of the electron density in the t_{2g} orbitals of Fe^{2+} (note that all of these chelates are low-spin, reflecting the high ligand field strengths of their ligands) is transferred into the "LUMO", i.e., the lowest unoccupied molecular orbital, of the ligand. It is generally recognized that bipy and phen are not very strong σ-type donors, but are capable of forming π-type coordinate bonds in which the electrons in the t_{2g} orbitals of the metal are, to some extent, delocalized back into the π-electron system of the ligands (back coordination). In fact, this effect is considered to be the reason why such ligands often stabilize the lower oxidation states of transition metals (Fe(II) in the present case). One can, therefore, imagine that the π-bonding electrons of the excited system are delocalized even further into the LUMO of the ligands, resulting in a strong CT band in the visible region [7].

A number of mixed complexes of these ligands with CN^-, such as $[FeL_2(CN)_2]$ or $[FeL(CN)_4]^{2-}$, (L = bipy or phen), have been known for a long time. They are also low-spin complexes which are quite stable and strongly colored in aqueous solutions. In fact, the spectral similarity between $[FeL_3]^{2+}$ and these mixed complexes has already been pointed out by Yamasaki, who studied their absorption spectra for the first time [8].

Schilt [9, 10] found that these mixed chelates (not the $[FeL_3]^{2+}$-type chelates) are very solvatochromic, i.e., their colors strongly depend on the polarity (and acidity; cf. below) of the solvent used. Soukup and Schmid recently described a number of interesting demonstration experiments with $[Fe(phen)_2(CN)_2]$, which is very soluble in a large number of solvents [11, 11a]. According to them, a marvellous spectrum of colors is observed, when a small amount of this complex is added to each of the following solvents:

Solvent	HMPA	DMF	CH_2Cl_2	EtOH	CH_3COOH	HCOOH	CF_3COOH
Color	blue	bluish violet	violet	wine red	red	orange	yellow
(AN)	(10.6)	(16.0)	(20.4)	(37)	(53)	(84)	(105)

(Eq. E.3)

Soukup also pointed out that 3,4,7,8-tetramethyl-phen (abbreviated as tmphen) yields a corresponding complex, $[Fe(tmphen)_2(CN)_2]$, which is even more soluble in non-polar solvents. The colors of this complex observed in polar solvents are essentially similar to those of $[Fe(phen)_2(CN)_2]$. Hence, a comparative study of a larger number of solvents becomes possible [11].

It is fascinating that the color changes from blue to red and then yellow just in the order of AN, which is given below each solvent. This correlation with AN, which was pointed out by Mayer [12] after a foregoing study by Burgess, who instead used the related parameter E_T [13, 14], can now be explained as follows. Similar to bipy or phen, CN^- is also known to act as a strong π-acceptor ligand using its π* orbital. Therefore, when it shares the coordination sphere of Fe(II) with bipy or phen, it also competes with L (bipy or phen) for the t_{2g} electrons on Fe(II), as shown in Fig. E.3a.

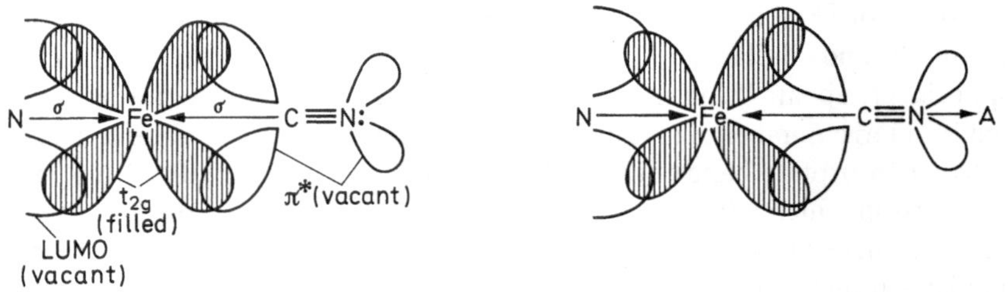

a

Fig. E-3 (a). Effect of an acceptor solvent (A) on the charge transfer in $[FeL_2(CN)_2]$ or $[FeL(CN)_4]^{2-}$ (L = bipy, phen and similar ligands). Left: overlap of a LUMO of L, a d-orbital (t_{2g}) of Fe^{2+} and a π* orbital of CN^- in the complex. Right: Electron withdrawal by A which weakens the charge transfer interaction from Fe^{2+} to L

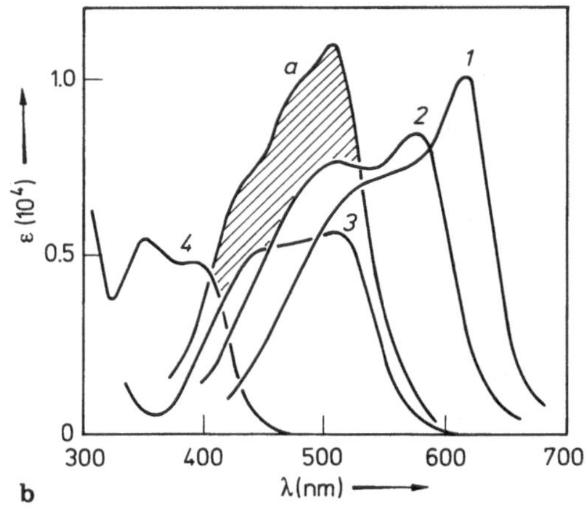

b

Fig. E-3 (b). Absorption spectra of $[Fe(phen)_2(CN)_2]$ (10^{-4} M) in DMF (*1*), CH_3NO_2 (*2*), 0.05 M HCl/H_2O (*3*), conc. H_2SO_4 (*4*). (*a*) is the spectrum of $[Fe(phen)_3]^{2+}$, which is quite insensitive to a change in solvent. (After Soukup and Schmid [11a])

When the complex is dissolved in an acceptor solvent, the lone pair of electrons protruding from the terminal nitrogen atom of the coordinated CN^- is easily solvated, as shown in the same figure. The electronic shift brought about by the solvation strengthens the π-bonding interaction between Fe^{2+} and CN^-, but weakens that between Fe^{2+} and L. Therefore, the $\tilde{\nu}_{max}$ value of the CT band of the "iron-methine chromophore" increases drastically with an increase in the AN of the solvent.

The spectral data given in Fig. E.3b confirm these expectations; in this case, the AN of 0.05 M HCl can be taken to be slightly higher than that of H_2O, i.e., ca. 60, while the AN of concentrated H_2SO_4 is estimated to be ca. 130 [11]. According to Soukup and Schmid [11a], who strongly recommend $[Fe(phen)_2(CN)_2]$ as an indicator of solvent AN (and also $[Cu(tmen)(acac)]ClO_4$ as an indicator of DN; cf. Chap. D.III.2), the relation $AN = -133.8 + 0.00933(\tilde{v}_{max})$ is valid for the spectra measured in 12 solvents.

Systematic studies on solutions in acids also revealed that species such as $[FeL_2(CN)(CNH)]^+$ and $[FeL_2(CNH)_2]^{2+}$ are formed successively in the solution, influencing the colors and spectra; in this case the acceptor is H^+ [9, 10]. Soukup and Schmid found another color change series: if $[Fe(phen)_2(CN)_2]$ is dissolved in CH_3CN and $NH_4B(Ph)_4$, $NaB(Ph)_4$, $Mg(ClO_4)_2$, $Fe(ClO_4)_2$, $Co(ClO_4)_2$, $Fe(ClO_4)_3$ and $Al(ClO_4)_3$ are each added to a test tube containing the solution, a spectrum of colors from violet over red to yellow is again observed. In this case the acceptor is NH_4^+ or one of the metal ions, which forms a coordinate bond with the nitrogen end of the CN^- ion. The strength of the coordinate bond increases from $NH_4B(Ph)_4$ to $Al(ClO_4)_3$ [11, 11a].

A recent paper by Toma and Takasugi [15] is interesting in this connection. They carried out a large number of spectral measurements with the three types of complexes, $[Fe(L)_2(CN)_2]$, $[Fe(L)(CN)_4]^{2-}$ and $[FeY(CN)_5]^{3-}$ (L = bipy, phen and Schiff base derivatives of pyridine; Y = pyridine and related monodentate ligands) in solvents of widely different AN, as well as in mixed solvents. In each case, they observed a remarkable solvatochromism, which can be precisely correlated with the AN (cf. Fig. E.4).

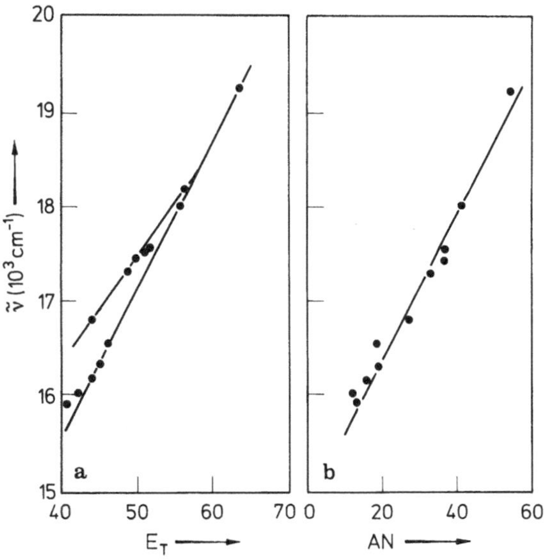

Fig. E-4 a, b. Relation between $\tilde{v}_{max}$ of $[Fe(bipy)_2(CN)_2]$ and E_T **(a)** and AN **(b)**. From top to bottom, the points correspond to H_2O, glycol (lacking in **(b)**), MeOH, EtOH, n-PrOH (overlapping with EtOH in **(b)**), n-BuOH, iso-PrOH, tert-BuOH, CH_3CN, DMSO, DMF, acetone, and py, respectively. Note that the points in **(a)** lie on two straight lines, one for protic and another for aprotic solvents, but those in **(b)** lie on only one such line. (Taken from Toma and Takasugi [15])

They also pointed out the fact that, when a ligand such as pyrazine is used in the $[FeY(CN)_5]^{3-}$ series, the complex becomes much less solvatochromic than $[Fe(py)(CN)_5]^{3-}$, due to the solvation of the ligands as shown below:

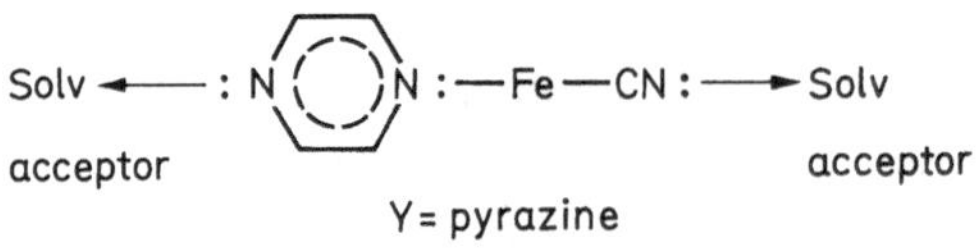

The effect of the solvation of Y, and that of CN^-, thus partially cancel out. The same effect was also observed with the Schiff base chelates, where a part of the chelate ring is solvated, counteracting the effect on CN^-. This kind of "asymmetric solvation", and "preferential solvation" which occurs in mixed solvents, can be very effectively investigated using these complexes, as Toma and Takasugi have skillfully tried to do.

Several other complexes are known to change color depending on the AN of the solvent; they include thiocyanato complexes such as $[Cr(NCS)_6]^{3-}$, and metal carbonyls such as $[Mo(bipy)(CO)_4]$ or $[W(bipy)(CO)_4]$. The mechanism of their color changes is expected to be similar to that of the Fe(II) complexes described above, since each of their linear π-acceptor ligands has a lone pair of electrons at its noncoordinated end, which is exposed to the easy attack of acceptor solvents [13, 14a]. Piezochromic and thermochromic behavior of some of these complexes were also studied [15a].

Fe(III) cyanide-(heterocyclic N-base) complexes are also deeply colored, and show a solvatochromism which is just the reverse of the Fe(II) complexes. For example, $[Fe(phen)_2(CN)_2]^+$ is red in CH_3NO_2, reddish violet in H_2O, and blue to green in strong acids. If the origin of the colors of these complexes is a CT band of ($M \leftarrow L$) type, this trend seems natural; but it seems to be a d-d band, and Soukup and Schmid explained the trend by assuming the weakening of the ligand field of CN^- by acceptor solvents [11a].

Another example of solvatochromism, in which the AN plays a determining role, is that of $[Co(edta)]^-$ reported by Taura [12a]. He found that the d-d band of this chelate at 536 nm in H_2O is shifted to the red in organic solvents, up to 547 nm in acetone with an apparent color change from reddish violet to bluish violet. There is a good linear relationship between λ_{max} and AN for the six solvents used. (In certain cases, a cyclic polyether was used to solubilize the chelate, but this was shown not to have any influence on the spectra.) The observed change in color will certainly be related to the decreased solvation on the COO^- groups in the chelate anion.

E.IV Solvatochromism of Oxovanadium(IV) β-Diketonates

Another solvatochromic phenomenon, which has been known for years, is that of $[VO(acac)_2]$ and similar β-diketonates of VO^{2+}. They were originally studied by Selbin [16, 17], and later by Gutmann and Mayer [18]. This blue compound is composed of square pyramidal molecules as shown in Fig. E.5 [19]:

When this chelate is dissolved in a donor solvent, it is easily converted into a 6-coordinate solvate, $[VO(acac)_2(Solv)]$, which becomes more and more octahedral in structure with increasing donor ability (i.e. DN) of the solvent. A color change to green and notable spectral changes are brought about by this solvation.

$$[VO(acac)_2] + Solv \rightarrow [VO(acac)_2(Solv)] \qquad \text{(Eq. E.4)}$$

The d-d absorption of this chelate, in which V(IV) is a $3d^1$ system, appears in the red part of the visible spectrum in non-polar solvents. It is composed of two bands, I and II; the former (with a low $\tilde{\nu}_{max}$) being stronger. With an increase in the DN of the solvent, band I is notably shifted to the red, while band II is shifted a little to the blue, so that the

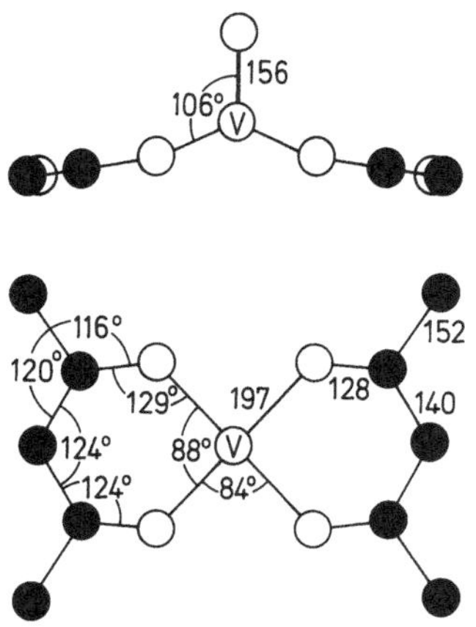

Fig. E-5. Top and side views of a $VO(acac)_2$ molecule. Bond lengths are in pm. Note that the V = O bond is much shorter than the V–O(acac) bonds, and that the two chelate rings are bound to V^{IV} like the two wings of a bird. (After Dodge et al. [19])

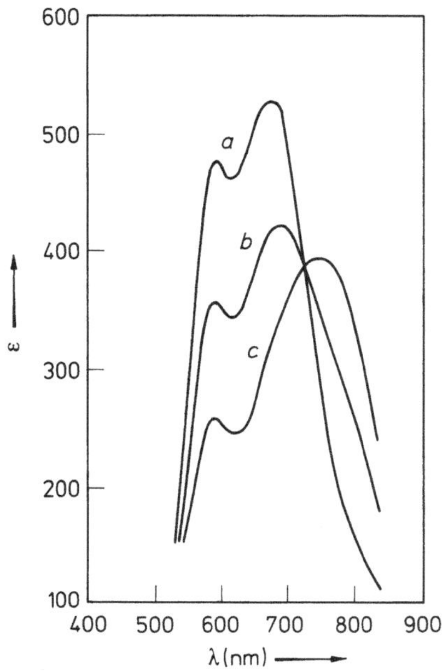

Fig. E-6. Absorption spectra of $VO(acac)_2$ in CH_2Cl_2 for increasing concentrations of DMSO. Conc. of $VO(acac)_2$: 1.08×10^{-2} M; conc. of DMSO: *a*, 0; *b*, 0.122 M; *c*, 0.388 M. (After Gutmann and Mayer [18])

difference in their $\tilde{\nu}_{max}$ values ($D_{II,I}$) increases strongly with the DN of the solvent used. This same change can be brought about by diluting a solution with a solvent of higher DN, as can be seen from Fig. E.6 [18].

Gutmann and Mayer [18] obtained many such curves. They determined the equilibrium constants of (Eq. E.4), by adding a strong donor solvent such as py, HMPA and DMSO to the chelate dissolved in a much weaker donor solvent such as CH_2Cl_2 or CH_3CN. They also tried to measure the donor ability of various anions, by using their salts instead of the strong donor solvents. Their results show that the stability of $[VO(acac)_2(Solv\ or\ Anion)]^{0\ or\ -}$ changes in the following order:

Ligand: $N_3^- > NCS^- > HMPA > py > Ph_3PO > DMSO > Cl^- > DMF > Me_3PO > Br^- > I^-$.
(DN) (38.8) (33) (29.8) (26.6) (23.0)
(Eq. E.5)

As expected, the order for the solvents is that of their DN. The order for the anions is:

$$N_3^- > NCS^- > Cl^- > Br^- > I^-; \quad \text{(Eq. E.6)}$$

Soukup [20] added two more anions, $OH^- > CN^-$, to the left-hand end of Eq. E.6, and pointed out that the entire order is essentially the same as that of Edward's H-value. This order can be taken to be a measure of their nucleophilicities [21].

If the value of $D_{II,I}$ is plotted against the DN of the solvent, a good linear relation is obtained for most aprotic donor solvents [22]. Therefore, $D_{II,I}$ can serve as a new polarity parameter for these solvents. However, protic solvents, which can form hydrogen bonds with anions or the negative atoms of the solutes (and therefore exhibit very high AN values), generally yield $D_{II,I}$ values which are much higher than those expected from their DN alone. Gutmann [22] ascribed this effect to the solvation of the oxygen atom of the V=0 group by solvent molecules which behave as acceptors.

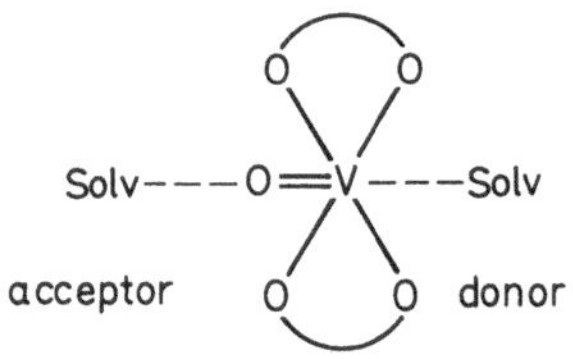

It is easy to see that such a solvation weakens the V=0 bond, making it more polar (i.e., like $V^+—O^-$). This, in turn, favors a solvation by donor-like solvent molecules from behind the plane of the chelate. It is thus conceivable that such a "push-and-pull" of electrons leads to a much stronger solvation in protic solvents.

It is interesting to note that there is a correlation between $D_{II,I}$ and AN for these protic solvents, as can be seen below:

	H_2O	$HCONH_2$	MeOH	EtOH	n-BuOH
$D_{II,I}$	42.3	39.1	33.8	30.4	29.0
AN	54.8	39.8	41.5	37.9	36.8
DN	16.4	24	19.1	----	----

It is tempting to express these $D_{II,I}$ values as a function of a(AN) + b(DN), as was done for the mixed chelates of Cu(II) (cf. Chap. D). The lack of sufficient DN values, however, hinders this at present.

Similar solvatochromism is also observed with other β-diketonato chelates of VO^{2+}. Comparative studies by Adachi et al. [23], who also carried out some thermochromic studies, indicate that such a solvatochromism can best be observed with $[VO(acac)_2]$ and $[VO(bzac)_2]$. $[VO(dipm)_2]$ shows only a weaker tendency to combine with the solvent, so that its solvatochromism is only observed when the DN of the solvent is high. On the other hand, $[VO(tfac)_2]$ easily combines with most solvents, and, in solvents of low DN, it tends to polymerize as follows:

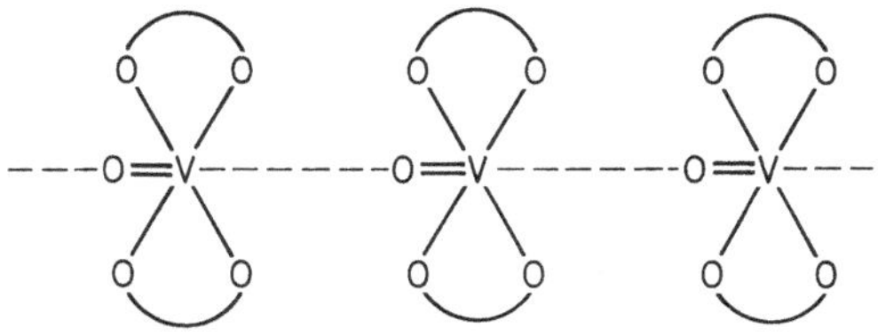

It is, therefore, almost always 6-coordinate, so that its solvatochromism is not very distinct.

There are certainly many things to do with these and related chelates of VO^{2+}, both in theory and practice just like those of Ni^{2+} and Cu^{2+} treated in Chaps. C and D.

E.V Thermochromism of Metal Chelates with Triphenylmethane Complexones in Aqueous Solutions

The chromotropic changes of metallic chelates that have been treated so far have involved changes in the d-d absorption, or in the CT bands, caused by structural and electronic changes in solution.

A different type of thermochromism, in which the color changes are due to changes in the absorption of the π-electron systems of the ligands, was observed by Fujimoto and his collaborators [24–28]. Everyone knows that metal ions can be conveniently titrated with edta, using special indicators (metal indicators) that are chelating agents in themselves. The strongly colored chelates formed by such indicators are often remarkably thermochromic. For example, the orange color of a solution containing XO (xylenol orange; see below) and a large excess of Cu^{2+} at pH 4 becomes reddish violet at 60 °C. Figure E.7 shows the accompanying spectral changes.

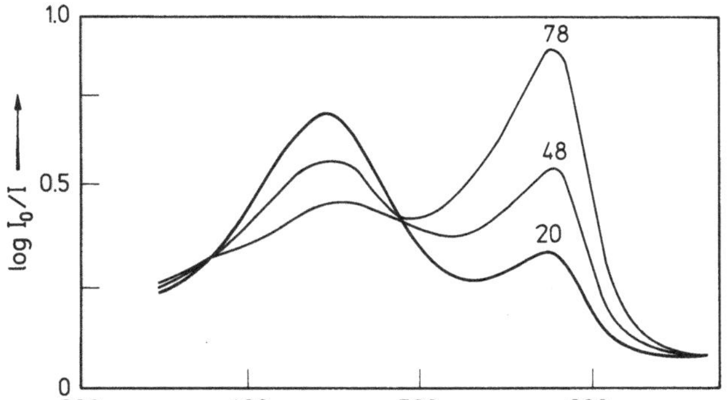

Fig. E-7. Thermochromism of a solution containing 1.2×10^{-4} M Cu^{2+} and 3.0×10^{-5} M XO at pH 3.94 and I = 0.1 M. The numbers on the curves are temperatures (°C). (After Nakada et al. [27])

Nakada et al. [26–28] listed more than 30 such examples, using 10 indicators of the "triphenylmethane complexone" type and bivalent cations such as Mg^{2+}, Ca^{2+}, Mn^{2+}, Co^{2+}, Ni^{2+}, Cu^{2+} and Zn^{2+}. In most cases, the observed color changes are from yellow or orange (sometimes even colorless) at room temperature to violet, green or blue at 60 °C. In each case, they are completely reversible. When complexometric titrations with these indicators are carried out in hot solutions, these thermochromic changes make it possible to determine Co^{2+}, Ni^{2+}, Cu^{2+} and Zn^{2+} at low pH's. Such titrations were previously impossible [25].

Nakada et al. proposed the following mechanism for the color change. For example, in the case of the Cu^{2+}—XO system, which forms a 2:1 chelate in an excess of Cu^{2+}, one can easily understand that each of the two Cu^{2+} ions in the chelate is held by each of the —$CH_2N(CH_2COO)_2$ groups.

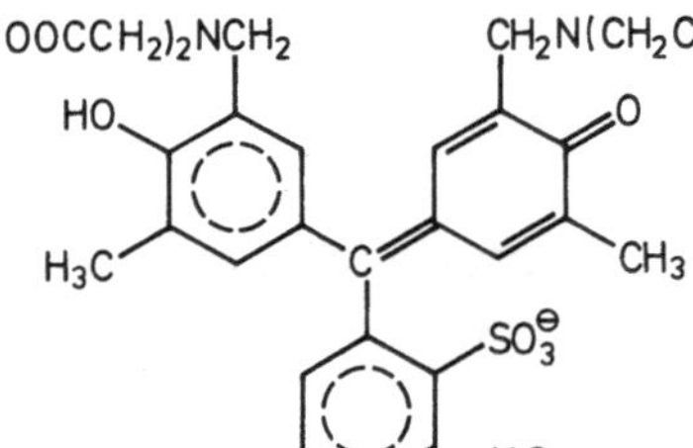

XO

The phenolic OH group, which is in the vicinity of a Cu^{2+} ion, seems to be free in the orange species, but to get ionized and coordinated to Cu^{2+} in the violet species. The thermochromic equilibrium can thus be described as a protonation-deprotonation coordination equilibrium of the ligand. A change in the π-electron system brings about the spectacular change in color.

E.VI Thermochromism and Solvatochromism Accompanied by the Formation of Ligand Radicals in Solution

Recently, Chavan et al. [29] found that the Co(III) and Ni(III) complexes of the ligands L show a new type of thermochromic and solvatochromic behavior.

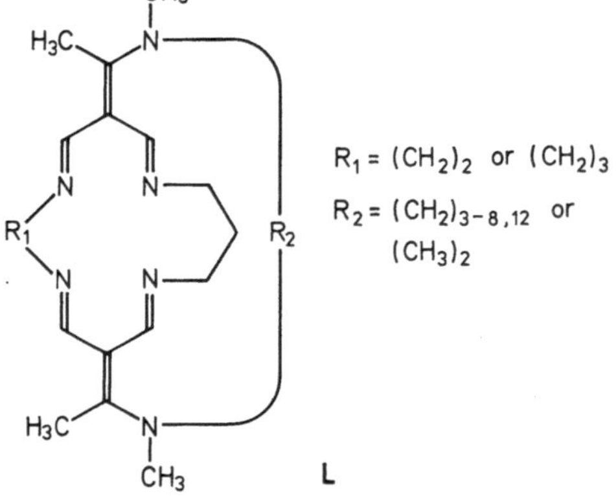

L

These complexes are obtained by electrolytic oxidation of the Co(II) and Ni(II) complexes in organic solutions. For example, the Ni(III) complex of L ($R_1 = (CH_2)_3$ and $R_2 = (CH_3)_2$) in CH_3CN is green, but its absorption band at 590 nm disappears on cooling to ca. −30 °C, leaving a broad band in near infrared. ESR study indicates that this is due to a shift of an equilibrium between the solvated Ni(III) complex and the coexisting Ni(II)-(ligand radical) complex to the left hand side:

$$[Ni^{III}L(Solv)_2]^{3+} \rightleftharpoons [Ni^{II}(\cdot L^+)]^{3+} + 2Solv$$

Similar equilibria were also found in the case of the other complexes. The dependence of the proportions of Co(III)L and Co(II)($\cdot L^+$) complexes on the nature of the solvent (solvatochromism) was also indicated.

These phenomena are clearly of interest in connection with metal-catalyzed redox reactions in biological systems.

References

1. Cf. e.g. Goodwin, H. A.: Coord. Chem. Rev. **18**, 293 (1976)
2. Chum, H. L. et al.: Inorg. Chem. **21**, 1146 (1982)
3. Matsumoto, N. et al.: Chem. Lett. **1984**, 479

3a. For a study on binuclear Fe(III) complexes, see: Ohta, S. et al.: Bull. Chem. Soc. Jpn. **59**, 155 (1986)

4. Sone, K.: Bull. Chem. Soc. Jpn. **25**, 1 (1952)
5. Krumholz, P.: J. Am. Chem. Soc. **75**, 2163 (1953)
6. Busch, D. H., Bailer, J. C. Jr.: Am. Chem. Soc. **78**, 1137 (1956)
7. Krumholz, P.: "Structure and Bonding", Vol. 9, p. 139 (1971)
8. Yamasaki, K.: Bull. Chem. Soc. Jpn. **15**, 461 (1940)
9. Schilt, A. A.: J. Am. Chem. Soc. **82**, 3000 (1960)
10. Schilt, A. A.: ibid. **82**, 5779 (1960)
11. Soukup, R. W.: Chem. in unserer Zeit **17**, 129 (1983)

11a. Soukup, R. A., Schmid, R.: J. Chem. Educ. **62**, 459 (1985)

12. Mayer, U.: Pure Appl. Chem. **51**, 1697 (1979)

12a. Taura, T.: Chem. Lett. **1984**, 2011

13. Burgess, J.: Spectrochim. Acta **26A**, 1369 (1970)
14. Burgess, J.: ibid. **26A**, 1957 (1970)

14a. Burgess, J.: J. Organometall. Chem. **19**, 218 (1969)

15. Toma, H. E., Takasugi, M. S.: J. Soln. Chem. **12**, 547 (1983)

15a. Macholdt, H.-T. et al.: Inorg. Chim. Acta **104**, 115 (1985)

16. Selbin, J., Ortolano, T. R.: J. Inorg. Nucl. Chem. **26**, 37 (1964)
17. Selbin, J.: Chem. Rev. **65**, 168 (1965)
18. Gutmann, V., Mayer, U.: Monatsh. Chem. **99**, 1383 (1968)
19. Dodge, R. P. et al.: J. Chem. Phys. **35**, 55 (1961)
20. Soukup, R. W.: Chem. in unserer Zeit **17**, 163 (1983)
21. Ibne-Rosa, K. M.: J. Chem. Educ. **44**, 89 (1967)
22. Gutmann, V.: "Coordination Chemistry in Non-Aqueous Solutions", Springer, Wien (1968), p. 22
23. Adachi, N. et al.: Bull. Chem. Soc. Jpn. **50**, 401 (1977)
24. Nakada, S. et al.: Chem. Lett. **1977**, 1243
25. Yamada, M. et al.: ibid. **1978**, 1153
26. Nakada, S. et al.: Bull. Chem. Soc. Jpn. **52**, 766 (1979)
27. Nakada, S. et al.: ibid. **53**, 2252 (1980)
28. Nakada, S. et al.: ibid. **54**, 2913 (1981)
29. Chavan, M. Y. et al.: Inorg. Chem. **25**, 314 (1986)

CHAPTER F

Thermochromism of Transition Metal Complexes in the Solid State

F.I Introduction

In addition to the thermochromic phenomena observed in solution, numerous changes are also known to occur in the solid state among the transition metal complexes [1, 2]. In this Chapter, we shall first consider some representative examples of irreversible thermochromism. Common reactions of colored complexes will not be treated, since they are, quite naturally, irreversibly thermochromic. Thus, we shall focus our attention on certain reactions of special structural interest, e.g., dehydration reactions, isomerizations and so on, in order to find out some general features of the phenomena. Examples of reversible thermochromism, which are of more specific interest, will then be reviewed.

We can classify the phenomena of irreversible thermochromism into two general types. In one of them, the change is truly irreversible, i.e., there is no equilibrium between the reactants and products during the reactions which are thermodynamically favored at the temperature of change. This "thermodynamically irreversible thermochromism" is most common among the reactions of colored complexes. The other type of irreversible thermochromism, termed "kinetically irreversible thermochromism", involves a reversible equilibrium; the reaction which occurs upon cooling the heated solid is, however, so slow that the system remains in its high-temperature state for a long time. Or, vice versa, the low-temperature state of the system is "deeply frozen"; here the reaction velocity is so slow that only strong heating can wake up the system. In these cases, we only observe the colors of metastable systems.

The phenomena of reversible thermochromism can also be classified into two types: (1) "continuous thermochromism", in which the change in color takes place continuously over a relatively wide range of temperature, corresponding to a gradual structural modification in the solid, and (2) "discontinuous thermochromism", which occurs more or less abruptly at a fairly distinct "thermochromic color-change point (T_{th})", corresponding to an abrupt change in structure.

These changes are reversible, but a certain amount of hysteresis may be observed. Hence, the T_{th} values observed upon cooling may be quite different from those observed upon heating. When this hysteresis is large enough, the phenomenon become kinetically irreversible.

Most of the examples discussed below can be reasonably classified into one of these categories, although a number of cases remain ambiguous. Spectral changes accompanying these phenomena usually involve characteristic shifts and deformations of the d-d bands, caused by the changes in the strength and symmetry of the ligand field, and often those of the CT bands due to changes in metal-ligand redox-type interactions. These situations can often be explained with ease, when the changes involve irreversible reactions, or discontinuously reversible structural changes. On the other hand, it should be noted that the positions and widths of these bands are influenced by the temperature, even in the absence of such drastic structural changes. Usually the d-d bands are shifted to the blue and sharpened (sometimes split into component bands) upon cooling, and vice versa. These changes are due to molecular vibrations; at lower temperatures, the weakening of such vibrations leads to a closer metal-ligand contact and more well-defined energies of the ground and exited states, bringing about the observed changes. In many cases "continuous thermochromism" is the result of this effect, although there are cases in which a real structural change takes place over a wide temperature range.

A reasonable understanding and a careful discrimination of the factors influencing spectral changes are thus important requisites for the study of solid-state thermochromism.

F.II Irreversible Thermochromism

F.II.1 Visual and Spectral Observations of the Color Changes

When colored transition metal complexes are heated slowly and carefully, it is often observed that they undergo spectacular color changes which sometimes occur in steps. Wilke and Opfermann [3] tried to observe this phenomenon with a number of Co(II) and Co(III) salts and complexes, by scattering powdered samples over a strip of aluminium (20 cm × 2 cm) along which a temperature gradient has been set up. Some of their results are shown in Table F.1.

Table F.1. Thermochromism of Co(II) and Co(III) complexes. The numbers are temperatures (°C); see text for the meaning of I, II, and III. (After Wilke and Opfermann [3]).

$CoSO_4 \cdot 7H_2O$:	Dark red $\xrightarrow[I]{60}$ Lilac $\xrightarrow[I]{170}$ Blue $\xrightarrow[I]{235}$ Dark red violet
$CoCO_3 \cdot 6H_2O$:	Violet $\xrightarrow[I,III]{162}$ Grey $\xrightarrow[III]{195}$ Black
$Co(OAc)_2 \cdot 4H_2O$:	Violet $\xrightarrow[I]{95}$ Bluish violet $\xrightarrow[I,III]{170}$ Black grey
$Co(C_2O_4) \cdot 2H_2O$:	Rose red $\xrightarrow[I]{150}$ Lilac $\xrightarrow[II]{230}$ Black
$CoSiF_6$:	Rose red $\xrightarrow[III]{105}$ Violet $\xrightarrow[III]{265}$ Black
$[Co(NCS)_2py_4]$:	Rose red $\xrightarrow[III]{85}$ Bluish green $\xrightarrow[III]{120}$ Black grey
$[Co(NH_3)_6]PO_4 \cdot 6H_2O$:	Red $\xrightarrow[I,III]{185}$ Violet $\xrightarrow[III]{285}$ Blue
$[Co(NH_3)_5Cl]Cl_2$:	Rose red $\xrightarrow[III]{120}$ Violet $\xrightarrow[III]{170}$ Turquoise $\xrightarrow[III]{230}$ Black

Wilke and Opfermann classified these changes into three categories, i.e., (I) release of crystal water (50–150 °C; mostly red violet → blue), (II) oxidation (> 200 °C, leading to black Co_3O_4), and (III) other kinds of thermal decomposition (e.g., release of CO_2, NH_3, etc.; various color changes), without going into the details of each. More data on salts of various metals were reported by Gvozdov and Erunova [4].

To study these color changes further, it is desirable to have a device to follow the spectral changes in solid samples during the course of heating or cooling. Wendlandt [5, 6, 61] devised a technique for this purpose and named it "dynamic reflectance spectroscopy (DRS)". He made use of a small, disk-shaped aluminium block (ϕ 6 cm, thickness 1.1 cm) to hold the powdered sample, which was packed into the coinlike hollow (ϕ 2.5 cm, depth 1 mm) on the top. The reflection spectra were then measured at various temperatures produced by a heating wire and monitored by a pair of thermocouples (Fig. F.1).

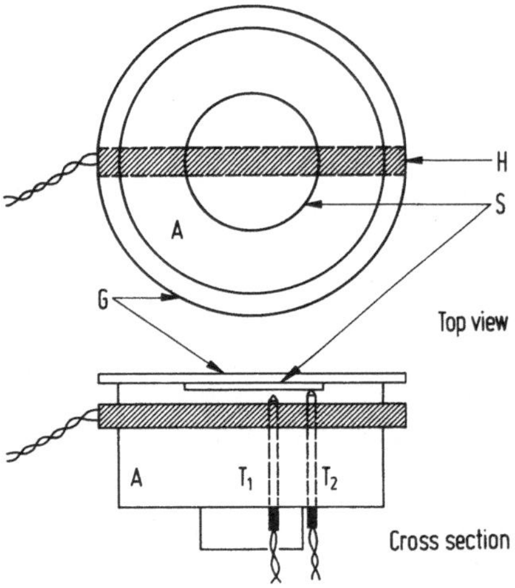

Fig. F-1. Heated sample holder used in "dynamic reflectance spectroscopy". A: Al block, H: heater, S: sample, G: glass cover, T_1, T_2: thermocouples to measure the temperature of H and S, respectively. (After Wendlandt and Smith [6])

In this way, Wendlandt studied the thermochromism of $CoCl_2.nH_2O$, $CoBr_2.6H_2O$, $Co(py)_2Cl_2$, $[Co(NH_3)_5H_2O]X_3$, $[Co(NH_3)_4(H_2O)_2]X_3$ ($X^- = Cl^-$, Br^-), $[Cu\,en(H_2O)_2]SO_4$, and many other complexes. For example, the data on $[Co(NH_3)_5H_2O]Cl_3$ are shown in Fig. 2a. It is evident that, above 75 °C, the spectrum of the sample changes completely from I (λ_{max}: 485 nm) to II (λ_{max}: 535 nm). Spectral and thermogravimetric(TG) studies have shown that this change is caused by the dehydration of the complex, which leads to the formation of $[Co(NH_3)_5Cl]Cl_2$ as the final product. The curve in Fig. F.2b reveals that the change takes place between 72–103 °C.

Multistep changes can also be followed in the same way. The data in Fig. F.3a and b show that $[Co(NH_3)_4(H_2O)_2]Cl_3$(I) is dehydrated in two steps, first forming $[Co(NH_3)_4(H_2O)Cl]Cl_2$(II) at 50–72 °C and then $[Co(NH_3)_4Cl_2]Cl$(III) (and some decomposition products) at 115–185 °C.

The usefulness of this spectral approach is evident. It is now common practice to follow solid state thermochromism in a similar way, using devices with various modifications, and often together with other thermochemical methods (TG–DTA

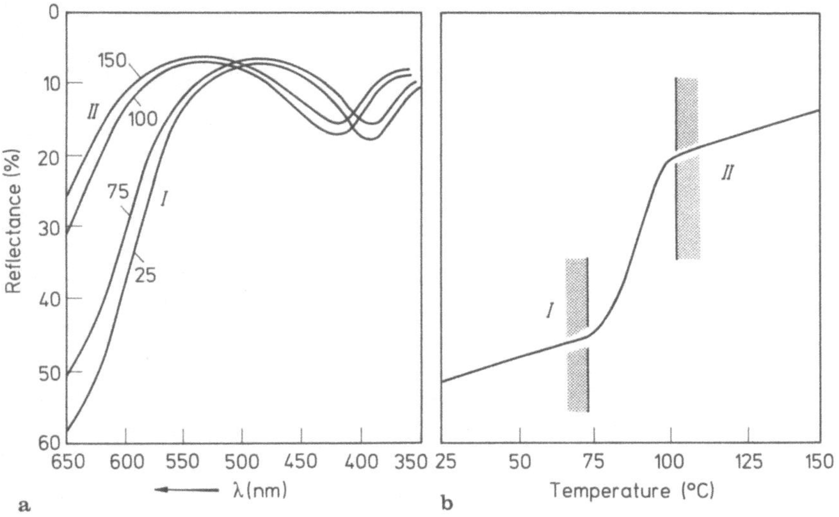

Fig. F-2 a, b. Reflection spectra of $[Co(NH_3)_5H_2O]Cl_3$ measured after heating **(a)**, and its % reflectance at 625 nm plotted against temperature **(b)**. The numbers on the curves are temperatures (°C). (After Wendlandt and Smith [6])

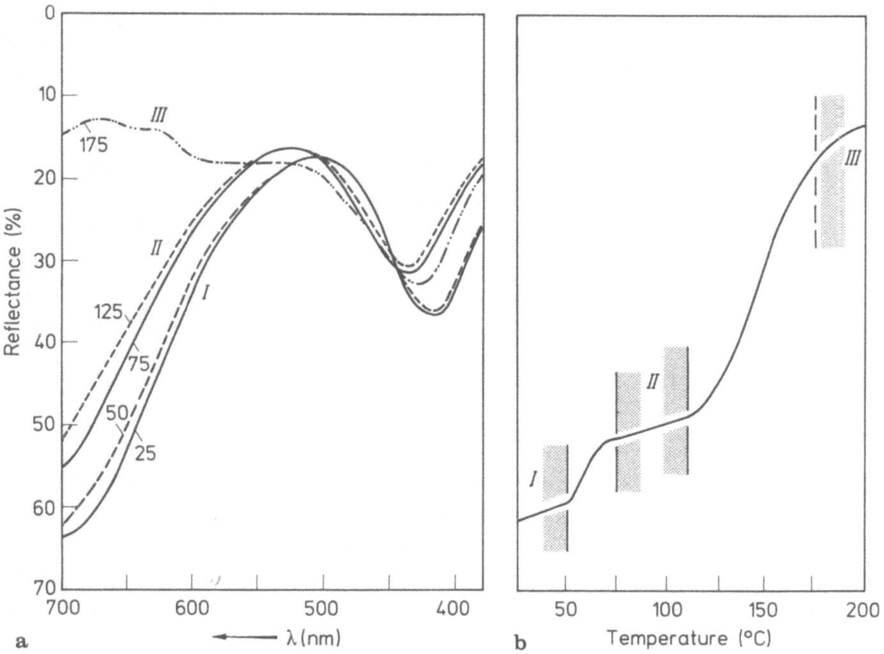

Fig. F-3 a, b. Reflection spectra of $[Co(NH_3)_4(H_2O)_2]Cl_3$ measured after heating **(a)**, and its % reflectance at 675 nm plotted against temperature **(b)**. The numbers on the curves are temperatures (°C). (After Wendlandt and Smith [6])

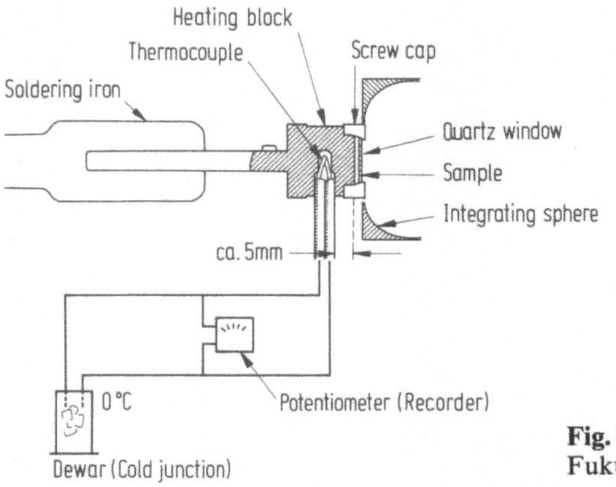

Fig. F-4. Heated sample holder used by Fukuda et al. (Taken from Fukuda [7a])

(differential thermal analysis) / DSC (differential scanning calorimetry)) and X-ray techniques to observe the nature of the structural changes involved. A sketch of the device that was used by Yamaki et al. [7] in their studies on $[Cu(daco)_2](NO_3)_2$ (cf. Chap. F.III.2.1) is shown in Fig. F.4.

F.II.2 Irreversible Thermochromism Caused by Dehydration

As can be seen from Table F.1, irreversible thermochromism due to dehydration, i.e., the release of coordinated water molecules, is very often observed when heavy metal salts and complexes are heated. The most drastic example of this phenomenon is shown by copper sulfate. It is well known that the blue crystals of $CuSO_4.5H_2O$ are readily converted into the white powder of anhydrous $CuSO_4$ upon heating gently. This powder, in turn, can serve as a dehydrating agent for organic solvents; absorbing traces of water from them, the blue color of $CuSO_4.5H_2O$ returns immediately.

One may ask: "Where does the blue color disappear upon heating? Where is the d-d band of Cu^{2+} in the spectrum of $CuSO_4$?" This problem was studied in some detail, making use of reflection spectroscopy and the TG-DTA technique. It is well known that this dehydration proceeds in a stepwise manner in the following way:

$$CuSO_4.5H_2O \xrightarrow{\text{ca. } 75\,°C} CuSO_4.3H_2O \xrightarrow{\text{ca. } 120\,°C} CuSO_4.H_2O \xrightarrow{\text{ca. } 240\,°C} CuSO_4 \quad \text{(Eq. F.1)}$$

Nagase et al. obtained solid reflectance spectra for each of these species and compared them with one another [8]. It was found that the d-d band is shifted only slightly in going from $CuSO_4.5H_2O$ to $CuSO_4$, i.e., from ca. 750 nm to ca. 800 nm, and a strong absorption, which is probably due to a $Cu^{2+} \leftarrow SO_4^{2-}$ CT band, appears in the near ultraviolet spectrum. Therefore, the disappearance of the blue color of $CuSO_4.5H_2O$ caused by dehydration is only an apparent phenomenon, which is brought about by a relatively small shift of the d-d band to the red end of the visible spectrum (and partially into the near infrared) and perhaps also by the spontaneous pulverization of the dehydrated product; if we had eyes which could register radiation from near IR to near

UV, we should certainly feel that $CuSO_4$ is much more darkly colored than $CuSO_4.5H_2O$!

It may be added that these results are compatible with the crystal structures of $CuSO_4.5H_2O$, $CuSO_4.3H_2O$ and $CuSO_4$ determined by X-ray analyses. Each Cu^{2+} ion in $CuSO_4.5H_2O$ is known to be surrounded by $4H_2O$ molecules, each at the corner of a square (Cu—O = 198 pm), and two SO_4^{2-} ions above and below the square (Cu—O = 238 pm). The result is a strongly elongated octahedron. In the course of dehydration, the water molecules in the coordination sphere are substituted, step by step, by the oxygen atoms of the SO_4^{2-} ions. In anhydrous $CuSO_4$, each Cu^{2+} ion is surrounded by sulfate ions which form an elongated octahedron as shown below.

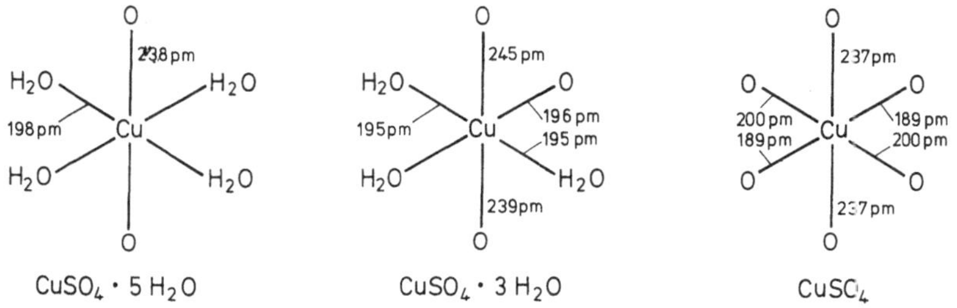

Since the ligand field strengths of H_2O and SO_4^{2-} are seemingly only slightly different from each other, these changes will only bring about a small shift in the d-d band towards the red. On the other hand, the SO_4^{2-} ions, which are weakly held in the axial positions of $CuSO_4.5H_2O$, are gradually brought into the equatorial coordination sphere, so that their CT interaction with the Cu^{2+} ion will be strongly enhanced. The observed spectral changes thus come about.

Somewhat similar spectral changes are also observed when the double sulfates $M_2Cu(SO_4)_2.nH_2O$ ($M = Na^+$, K^+, Rb^+, Cs^+ and NH_4^+) are heated. The double sulfate of $CoSO_4$ loses its water as follows:

$$\underset{\text{dark pink}}{Co_2Cu(SO_4)_3.18H_2O} \xrightarrow{32-76^\circ C} \underset{\text{pink}}{Co_2Cu(SO_4)_3.3H_2O} \xrightarrow{200-300^\circ C} \underset{\text{violet}}{Co_2Cu(SO_4)_3} \qquad \text{(Eq. F.2)}$$

Its spectral changes were compared with those of $CuSO_4.5H_2O$, and with those of $CoSO_4.6H_2O$ which take place as follows:

$$\underset{\text{reddish pink}}{CoSO_4.6H_2O} \xrightarrow{40-120^\circ C} \underset{\text{pink}}{CoSO_4.H_2O} \xrightarrow{250-300^\circ C} \underset{\text{violet}}{CoSO_4} \qquad \text{(Eq. F.3)}$$

It was found that, while the dark pink color of $Co_2Cu(SO_4)_3.18H_2O$ is essentially a superposition of the colors of the $[CuO_6]$ and $[CoO_6]$ chromophores, the colors of the trihydrate and anhydrous salts are only due to the $[CoO_6]$ chromophore because the CuO_6 chromophore becomes scarcely visible after dehydration (as in the case of $CuSO_4.5H_2O$). A strong CT absorption again appears in the ultraviolet region. In view of these data, it seems that the thermochromic behavior of other double salts containing two kinds of colored metallic ions should be an interesting topic for further studies.

Other examples of "dehydration thermochromism" can be found among the cobalt(II) and nickel(II) complexes of hexamethylenetetramine (hmta; Fig. F.5). These interesting compounds, which were discovered early in this century by Barbieri [10], are truly fantastic and seemingly capricious in their color changes. For example, $Co(hmta)_2Cl_2.10H_2O$ and $Ni(hmta)_2Cl_2.10H_2O$ are pink and green, respectively, which correspond to the colors of the hydrated ions of Co^{2+} and Ni^{2+}. When heated to ca. 150 °C the former chelate turns blue, while the latter at first turns yellow and then finally violet. Other halides, nitrates and thiocyanates, of the general composition $Co(hmta)_2X_2.nH_2O$ and $Ni(hmta)_2X_2.nH_2O$, also show characteristic color changes, depending on the nature of X and the number of H_2O molecules.

The structures of these complexes and the modes of their changes upon heating have been a mystery untill recently. Nagase et al. [11] studied theses complexes using various techniques involving TG-DTA, powder X-ray diffraction, reflectance and IR spectra. The results are the schemes shown in Table F.2. The reflectance spectra of $Ni(hmta)_2X_2.10H_2O$ and its decomposition products are shown in Fig. F.6.

The data show that the color changes are the results of progressive dehydrations. Each decrease in the number of H_2O molecules about a metal ion causes a gap in the coordination sphere, which is first filled by the anions X^-. The hmta molecules, which are very bulky and weak bases, are coordinated in the final stage, usually forming

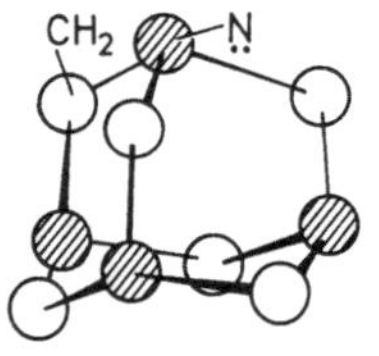

Fig. F-5. Molecular shape of hmta

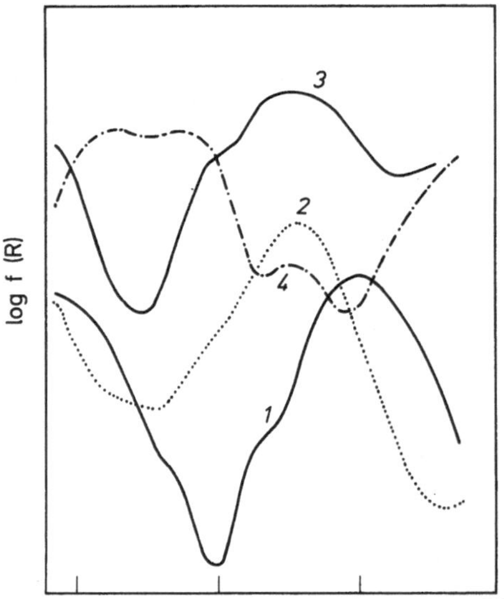

Fig. F-6. Thermochromic spectral changes of $NiCl_2(hmta)_2.10H_2O$. The reflection spectrum at room temperature (*1*) is very similar to the spectrum of an aqueous solution containing $[Ni(H_2O)_6]^{2+}$, while that of the dihydrate obtained at 110 °C (*2*) is similar to that of $NiCl_2.2H_2O$ (*3*). The anhydrous complex obtained at 150 °C shows a very different spectrum (*4*), indicating a tetrahedral structure. (Taken from Nagase et al. [11])

Table F.2. Color and structural changes in $Co(hmta)_2X_2.nH_2O$ and $Ni(hmta)_2X_2.nH_2O$ caused by heating. (Taken from Nagase et al. [11]). (L = hmta)

$[Co(H_2O)_6]Cl_2 \cdot 4H_2O \cdot 2L \xrightarrow{50-130°C} [CoCl_2L_2]^a$
Pink → Blue

$[Co(H_2O)_6]Br_2 \cdot 3H_2O \cdot 2L \xrightarrow{55-105°C} [CoBr_2L_2]^a$
Pink → Blue

$[Co(H_2O)_6]I_2 \cdot 2H_2O \cdot 2L \xrightarrow{55-110°C} [CoI_2L_2]^a$
Pink → Green

$[Co(H_2O)_6](NO_3)_2 \cdot 4H_2O \cdot 2L \xrightarrow{50-100°C} [Co(NO_3)_2(H_2O)_4] \cdot 2L \rightarrow$
Pink → Pink

$\xrightarrow{100-115°C} [Co(NO_3)_2(H_2O)_2] \cdot 2L^b \xrightarrow{125-145°C} [Co(NO_3)_2L_2]^b$
Violet → Violet

$[Co(NCS)_2(H_2O)_4] \cdot 2L \xrightarrow{85-125°C} [Co(NCS)_2L_2]^a$
Orange → Blue violet

$[Ni(H_2O)_6]Cl_2 \cdot 4H_2O \cdot 2L \xrightarrow{55-110°C} [NiCl_2(H_2O)_2] \cdot 2L^c \xrightarrow{120-155°C}$
Blue green → Yellow

$\rightarrow [NiCl_2L_2]^a$
Violet

$[Ni(H_2O)_6]Br_2 \cdot 3H_2O \cdot 2L \xrightarrow{70-125°C} [NiBr_2L_2]^a$
Blue green → Blue

$[Ni(H_2O)_6]I_2 \cdot 2H_2O \cdot 2L \xrightarrow{60-130°C} [NiI_2L_2]^a$
Blue green → Green

$[Ni(H_2O)_6](NO_3)_2 \cdot 4H_2O \cdot 2L \xrightarrow{45-100°C} [Ni(NO_3)_2(H_2O)_4] \cdot 2L \xrightarrow{100-130°C}$
Blue green → Yellow green

$\rightarrow$ (1.5 hydrate) $\xrightarrow{130-160°C} [Ni(NO_3)_2L_2]^b$
Yellow green → Green

[a] Tetrahedral.
[b] NO_3^-: probably bidentate.
[c] $[NiCl_2(H_2O)_2]$ in this complex is probably of an octahedral polymeric structure, such as:

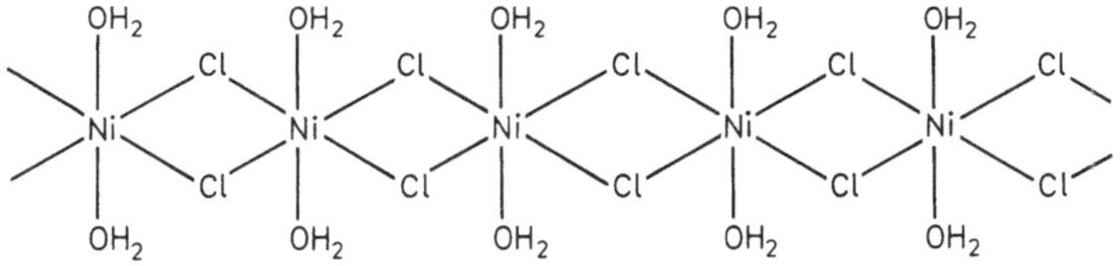

tetrahedral complexes as the final products. The features of the observed spectral changes can be quite reasonably understood making these assumptions.

The examples of dehydration thermochromism treated in this and next sections (Chaps. F.II.2–3) are not strictly irreversible; they are the shifts of "dehydration equilibria", which depend on temperature and water vapor pressure. However, their reverse changes ("hydration thermochromism") are often slow, making them look quite irreversible, especially when observed in dry atmosphere. They are, therefore, examples of "kinetically irreversible thermochromism" (cf. Chap. F.I).

F.II.3 Diamine Complexes of Nickel(II): Octahedral-Square Planar Changes with Deaquation and Anation

It was mentioned in Chap. C that $[Ni(en)_2]^{2+}$, and similar chelates with various alkyl or phenyl derivatives of en, exhibit interesting thermochromic behavior in solution. A systematic study of the solid state thermochromism of these chelates has recently been carried out by Ihara et al. [12]. Using N-alkyl derivatives of en(diam), they prepared the chelates $[Ni(diam)_2(H_2O)_2]X_2$, and obtained TG-DSC and spectral data on their thermal changes, some of which are shown in Fig. F.7. Comparing such data and the IR spectra and magnetic moments of the chelates and their heated products, they concluded that heating either brings about a "deaquation-anation", i.e., a release of coordinated water and a subsequent coordination by the anions, with retention of the coordination number 6 (case A):

$$\underset{\text{blue violet (n=0 or 2)}}{[Ni(diam)_2(H_2O)_2]X_2.nH_2O} \rightarrow \underset{\text{blue or green}}{[Ni(diam)_2X_2]} + (n+2)H_2O, \quad \text{(Eq. F.4)}$$

or a simple deaquation with a decrease in coordination number from 6 to 4, which changes the chelate into a yellow low-spin form (case B):

$$\underset{\text{blue violet}}{[Ni(diam)_2(H_2O)_2]X_2.nH_2O} \rightarrow \underset{\text{yellow}}{[Ni(diam)_2]X_2} + (n+2)H_2O. \quad \text{(Eq. F.5)}$$

The result of heating depends on the nature of diam and X. In some cases, the deaquation occurs spontaneously without heating, so that only the yellow, anhydrous form of the chelate is obtained (case C).

Table F.3A summarizes these three cases. In general, deaquation-anation predominates with the sym-diamine chelates, in which the axial coordination of anions is not

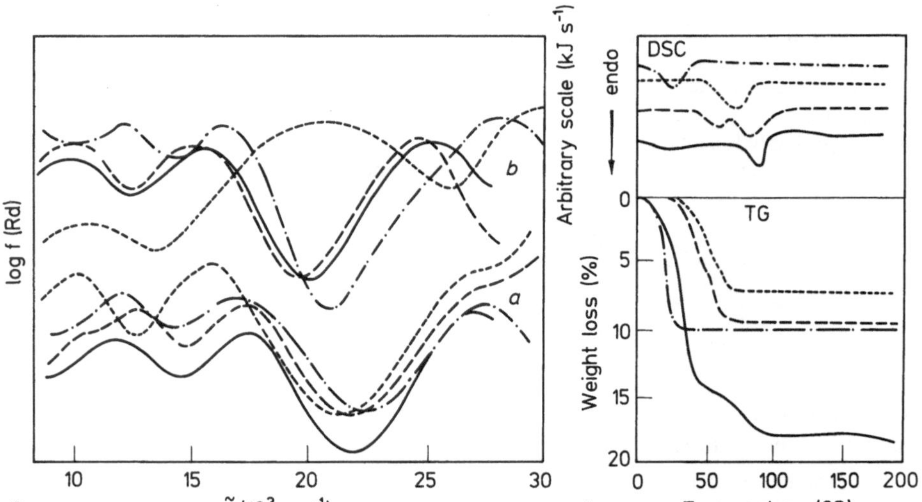

Fig. F-7 a, b. Reflection spectra of $[Ni(unsym\text{-}dmen)_2(H_2O)_2]X_2$ before (*a*) and after (*b*) heating (**a**), and the DSC and TG curves of the corresponding compounds (**b**). $X^- = Cl^-$ (with $2H_2O$, ——), Br^- (-----), I^- (·····), and NO_3^- (-·-·-·-·). (Taken from Ihara et al. [12])

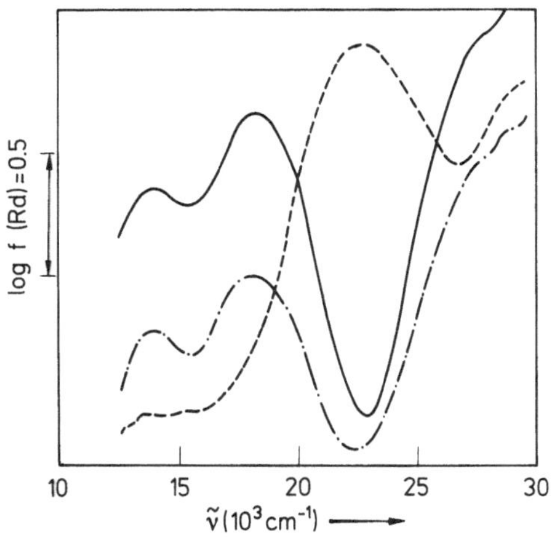

Fig. F-8. Reflection spectra of [Ni(d,l-stien)$_2$ $(H_2O)_2$]Br_2 before heating (——), and after heating at 120 °C (-----) and at 170 °C (-·-·-·-). (Taken from Ihara and Tsuchiya [13b])

sterically hindered (cf. Fig. C.11b). With unsym-dmen chelates, in which two methyl groups protrude axially above and below the plane of the chelate, the coordination of large anions such as I^- is hindered by such groups. Anation is hindered more strongly with unsym-deen chelates, so that only the Cl^- ion can enter the coordination sphere. Enthalpy changes for some of these changes were determined calorimetrically, furnishing additional evidence for such discussions [13].

Ihara et al. extended their studies to the same type of chelates with C-substituted ethylenediamines [13a–e]. Again the occurrence of the three cases A, B and C mentioned above can be recognized, but some of the chelates exhibit the following two-step change composed of "deaquation plus anation" (case D):

$$[Ni(diam)_2(H_2O)_2]X_2 \rightarrow [Ni(diam)_2]X_2 \rightarrow [Ni(diam)_2X_2]$$

blue violet — yellow — blue (Eq. F. 6a)

In most cases, the first change is completed below 100 °C, as observed with N-substituted diamines, while the second change occurs at 130–180 °C within a well-defined temperature range of 10–20 °C. These chelates exhibit a very fascinating thermochromism; when heated very gently, the violet crystals turn yellow, whereas when heated more strongly they again turn blue. The spectral data shown in Fig. F.8 clearly indicate that the pattern of the octahedral chelate, once completely lost by deaquation, returns via thermal anation. It was also observed that such a "thermal anation" can occur alone (case E):

$$[Ni(diam)_2]X_2 \rightarrow [Ni(diam)_2X_2] \qquad \text{(Eq. F.6b)}$$

yellow — blue

In such a case, the yellow chelate prepared in the usual way turns blue upon heating.

Table F.3 B summarizes these data. They can be understood by considering the competition between (i) the size and coordination ability of X^- and (ii) the steric hindrance of the substituent groups on diam, although the correlation is not as simple as in the case of N-substituted ligands. The anation steps in D and E are slightly endothermic (ΔH: ca. 10 kJ mol^{-1}). The ΔH values for the deaquation steps in B and D

Table F.3. Structural changes in $[Ni(diam)_2(H_2O)_2]X_2.nH_2O$ caused by heating (After Ihara et al. [12–13d]). Notations A–E mean that the following changes are observed (cf. also the corresponding equations in text):
A: Dehydration-anation (Eq. F.4),
B: Simple dehydration (Eq. F.5),
C: No reaction (always planar),
D: Dehydration plus anation (Eq. F.6a),
E: Simple anation (Eq. F.6b).

A: N-substituted ethylenediamine chelates					
diam	Cl^-	Br^-	I^-	NO_3^-	ClO_4^-
sym-dmen	A	A	A	A	–
unsym-dmen	A	A	B	A	–
sym-deen	A	A	A	A	A^a, B^b
unsym-deen	A	B	C	C	C

B: C-substituted ethylenediamine chelates					
diam[c]	Cl^-	Br^-	I^-	NO_3^-	ClO_4^-
pn	A	A	–	A	B
bn	D	D	–	A	B
dmbn	A	A	–	D	B
phenen	A	A	–	D	C
i-bn	E	C	C	C	C
m-bn	C	C	C	E	C
dl-bn	D	D	–	B	B
dl-peen	C	C	C	C	C
m-stien	C	C	C	C	C
dl-stien	D	D	C	D	C

[a] At 130 °C. [b] at 165 °C. [c] Substituent groups on the carbon atom(s) of en in these diamines are: 1Me(pn), 1Et(bn), 1tert-Bu(dmbn), 1Ph(phenen), 2Me on the same C(i-bn) and on different C's(m-bn and dl-bn), Et and Ph on the same C(dl-peen), 2Ph on different C's(m- and dl-stien), respectively.

are much higher (ca. 130 kJ mol^{-1}). Therefore, the color changes of the former steps are often notably reversible.

Thermochromic changes which are apparently similar to these were also observed by Tsuchiya et al. [14, 15], who studied the complexes of benzimidazole and 2-aminobenzimidazole (abbreviated as bimd and abi, respectively):

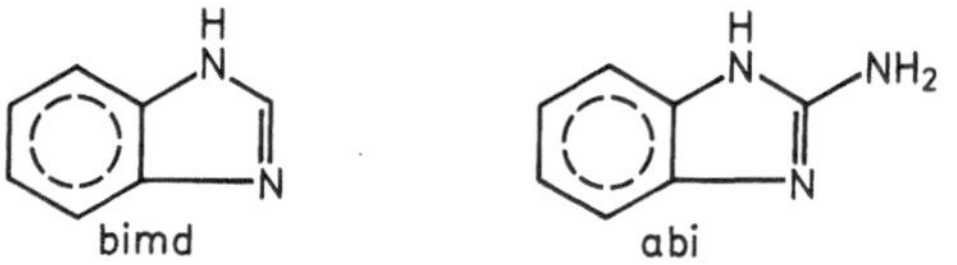

bimd abi

These complexes ($[Ni(bimd)_4](NO_3)_2.2.5EtOH$, $[Ni(bimd)_4]I_2$, and $[Ni(abi)_4]X_2.nH_2O$ ($X^- = Cl^-$, Br^- or NO_3^-; n = 1 or 3)) are all yellow and low-spin. They are converted into green, high-spin octahedral complexes according to the anation scheme given in Eq. F.6b. The ethanol or water molecules outside the coordination sphere are lost simultaneously. It is remarkable that the water in the abi complexes can only be lost at very high temperatures (usually above 150–200 °C). Crystal structure studies of these complexes will certainly be of interest.

In comparing the spectra of Ni(II) chelates of different stereochemistries, the work of Ferraro et al. [15a] on their piezochromism is of special interest. They prepared a large number of bi-, tri- and tetradentate chelates of Ni(II), containing N, P, As, Sb, O, S and Se as ligand atoms with various stereochemistries, and measured their visible spectra (d-d bands) under high pressures (up

4×10^4 kg·cm^{-2}). They found that 5-coordinate chelates of trigonal bipyramidal structures are strongly piezochromic, with band shifts towards the blue. They suggested that this effect may be used to distinguish this type of chelate from those of other geometries (octahedral, tetrahedral, square planar or square pyramidal), which are not as sensitive to compression. It is also interesting that the planar [Ni(dimethylglyoxime)$_2$] is a notable exception among the latter. It is strongly piezochromic with a band shift towards the red, as was previously observed by Zahner and Drickamer [15b].

F.II.4 Decomposition and Isomerization of Chromium(III) Complexes

Another series of thermochromic solid state reactions, which Tsuchiya has studied for many years, involves the thermal decomposition and isomerization of Cr(III) complexes. Some of his recent results will be reviewed here.

When $[Cr(en)_3]X_3$ ($X = Cl^-$ or NCS^-, yellow) is heated, it loses an en molecule to yield cis-$[CrCl_2(en)_2]Cl$ (red-violet) or trans-$[Cr(NCS)_2(en)_2]NCS$ (orange). Since the N atoms of each en molecule in $[Cr(en)_3]^{3+}$ occupy cis positions, the result with NCS^- is quite peculiar. Uehara et al. [16] prepared a large number of chelates $[Cr(aa)_3]X_3.nH_2O$, $[Cr(aa)_2(bb)]X_3.nH_2O$ and $[Cr(aa)(bb)(cc)]X_3.nH_2O$, where aa, bb, and cc represent different bidentate diamines such as en, pn and tn($=NH_2(CH_2)_3NH_2$). They confirmed that cis-$[CrX_2(diamine)_2]X$ is produced in each case upon heating when $X^- = Cl^-$, whereas the trans one is produced when $X^- = NCS^-$ (when $[CrCl_2tn_2]Cl$ and $[CrBr_2tn_2]Cl$ are prepared this way, the cis form obtained at first is transformed into the trans modification upon further heating). Moreover, the deamination temperatures of the NCS^--complexes are remarkably lower than those of the Cl^--complexes. It was also observed that, when two or three kinds of diamines are present in the coordination sphere, the one which is driven out by heating is the most volatile of them, i.e., the diamines are driven out in the order of en > pn > tn.

The reason for this difference between Cl^- and NCS^- was elucidated by Akabori and Kushi, who studied the crystal structure of optically active $(+)_{589}$-$[Cr(en)_3](NCS)_3$ and compared it with that of $[Cr(en)_3]Cl_3$ [17, 18]. The conformation of the latter complex is designated by the symbol $\Lambda(\delta\delta\delta)$. It is the most stable of the conceivable conformations. The cation is surrounded by Cl^- ions in a highly symmetrical way. In the former complex, a complicated network of hydrogen bonds exists between the N— and S— ends of the NCS^- ions and the —NH_2 groups of en, so that the cation is remarkably distorted, assuming a conformation $\Lambda(\delta\lambda\lambda)$ with somewhat higher energy. The arrangement of NCS^- ions about the cation is also strongly deformed. Two thirds of these anions are bound via their N-ends to the axial and equatorial N—H groups of the complex cation, while the remaining anions are only bound to the equatorial N—H groups. Moreover, a space close to each complex cation can be found where the packing of ions is noticeably loose. All of these factors seem to favour the escape of a single en molecule, and the combination of two NCS^- ions from opposite sides of the complex. In other words, a path for isomerization in the solid is already programmed into the crystal structure of the complex.

Tsuchiya et al. also studied the thermal cis-trans(violet-green) transformations of a large number of bis(diamine)chromium(III) complexes, $[MX_2(aa)_2]X.HX.nH_2O$ and $[MX_2(aa)(bb)]X.HX$, where aa and bb are en, pn, tn, and some of their C-alkyl derivatives, and $X^- = Cl^-$ or Br^- [19, 20]. They found that (i) trans $\rightarrow$ cis isomerizations take place when the complexes $[MX_2(aa)_2]HX.nH_2O$ with 5-membered chelate rings are heated; (ii) cis → trans isomerizations take place when the complexes with 6-membered chelate rings are heated; and (iii) only the trans → cis isomerization takes place with $[MX_2(aa)(bb)]X.HX$, even when one of the rings is a 6-membered ring. In each case, the release of H_2O and HX occurs first. The isomerization then takes place with activation energies of 150–190 kJ mol^{-1}. A "bond rupture" mechanism as shown below was proposed to account for these isomerizations:

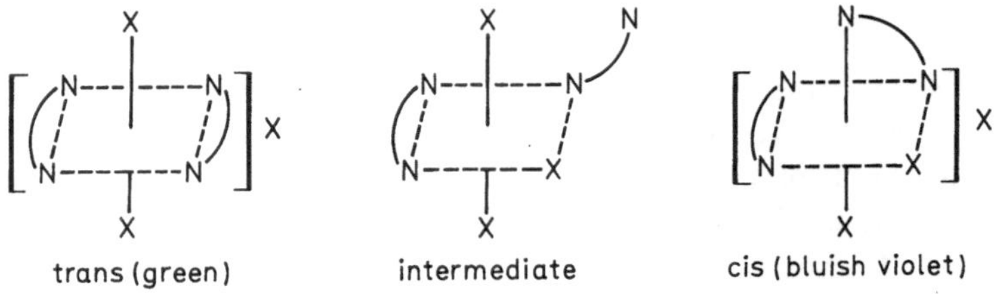

Ueno et al. also studied the thermal reactions:

$$[Cr(NH_3)_6]X_3 \longrightarrow [CrX_3(NH_3)_3] + 3NH_3 \quad \text{(Eq. F.8a)}$$

$$[CrX(NH_3)_5]Y_2 \longrightarrow [CrXY_2(NH_3)_3] + 2NH_3 \quad \text{(Eq. F.8b)}$$

$$\text{trans-}[CrX_2(NH_3)_4]Y \longrightarrow [CrX_2Y(NH_3)_3] + NH_3 \quad \text{(Eq. F.8c)}$$

where $X^-, Y^- = Cl^-$ or Br^- [21]. These reactions occur at relatively high temperatures, usually between 230–300 °C, and are notably thermochromic. Spectral and TG data offer interesting information concerning the replacement of ammonia by anions outside of the coordination sphere and the interaction of the latter with anions already present in the coordination sphere. The driving force for such a reaction is increased significantly when chelate-forming anions are present outside the coordination sphere. For example, with complexes such as $Na[Cr(NH_3)_6](edta).3H_2O$ and $K[Cr(NH_3)_6]Cl(nta).2H_2O$ (both yellow), the NH_3 and H_2O molecules are driven off upon heating slightly [22], and violet Na[Cr(edta)] and $K[Cr(nta)(NH_3)_2]Cl$ are obtained as products. It may be interesting to note that the observed color change indicates a weakening of the ligand field. Certain cis → trans isomerizations of Co(III) complexes were also studied by Tsuchiya et al. [19, 23].

F.III Reversible Thermochromism

F.III.1 Halide Complexes of Copper(II) and Nickel(II)

F.III.1.1 Thermochromism Associated with the Tetrahedral-Square Planar Changes of $[CuCl_4]^{2-}$ and Related Complexes

Among the complexes which exhibit a notable reversible thermochromism in the solid state, mention will first be made of the halide complexes of Cu(II) which have been extensively studied in recent years.

A large number of tetrahalogeno-complexes of Cu(II), $M^I_2CuX_4$ ($X^- = F^-, Cl^-$, or Br^-) are known. The fluoro complexes possess a polymerized structure containing octahedral $[CuF_6]$ units, whereas the chloro and bromo complexes contain discrete anions, $[CuX_4]^{2-}$, usually of a distorted tetrahedral structure. In the solid state, the degree of distortion not only depends on the nature of X^-, but also (and strongly) on the nature of M^+. When M^+ is an alkali metal cation, the $[CuCl_4]^{2-}$ anion is nearly tetrahedral. In $[Pt(NH_3)_4][CuCl_4]$, the same anion is nearly square planar. Various intermediate cases have also been observed, as in $Cs_2[CuCl_4]$ (< trans-Cl-Cu-Cl = $\alpha = 124°$) and $(C_{13}H_{19}N_2OS)_2[CuCl_4]$ ($\alpha = 143°$).

Willet et al. [24], who systematically studied the structures and spectra of these complexes, noted that the anions and cations are held together, not only by ionic bonds, but also by a network of N—H ... Cl hydrogen bonds extending over the entire crystal when the counter ion of $[CuCl_4]^{2-}$ contains N—H bonds. When this bond is strong enough, the negative charge on Cl^- is expected to be notably reduced. The usual stereochemistry about the Cu(II) ion, as discussed in Chap. C, is that of an elongated octahedron. The complex may become nearly square planar when the ligand field strength of the axial ligands is weak. Therefore, the $[CuCl_4]^{2-}$ ion should be intrinsically square planar. However, when the coordinated Cl^- ions retain most of their negative charge, their mutual repulsion forces the anion to assume a tetrahedral structure. This is the case for the alkali metal complexes, in which ionic interactions predominate. When the negative charge on Cl^- is reduced by hydrogen bond formation, the complexes tends to be more and more planar as the number and strength of the hydrogen bonds increases. Hence, one can estimate the presence and strength of these interactions by studying the shape of $[CuCl_4]^{2-}$.

Some of the tetrahalogeno complexes of Cu(II) were found to be strongly thermochromic. For example, when the green crystals of $[Et_2NH_2]_2[CuCl_4]$ are heated to 43° C, they abruptly turn yellow. The green color returns upon cooling to room temperature (this may take some time due to supercooling). DTA data have shown that this is due to an endothermic phase transition. The reflectance spectra in Fig. F.9 show a very drastic change within 2 °C.

A comparison of these curves with those of complexes of known structure indicates that the complex has nearly flat $[CuCl_4]^{2-}$ ions at room temperature. When heated, the ions assume a distorted tetrahedral structure ($\alpha = 133 \pm 5°$). This structural change, which is probably brought about by a weakening of the hydrogen bonds, is accompanied by a drastic change in the levels of the d-electrons. The complex, [iso-$PrNH_3]_2[CuCl_4]$, also shows the same type of thermochromism.

X-ray studies have confirmed this view, but have also shown that situation is not always that simple. For example: (i) The green, low-temperature form of $[Et_2NH_2]_2[CuCl_4]$ contains three kinds of anions exhibiting different degrees of flatness, while the yellow, high-temperature form

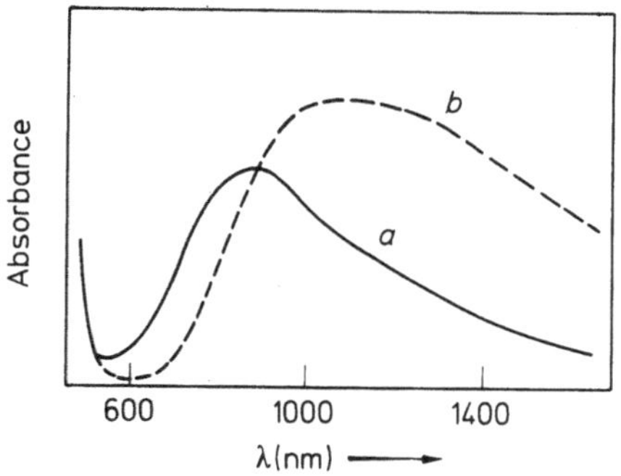

Fig. F-9. Thermochromic spectral changes of a sample of $[Et_2NH_2]_2[CuCl_4]$, sandwiched between glass plates, at 42 °C (*a*) and 44 °C (*b*). (Taken from Willett et al. [24])

contains two kinds of anions, both tetrahedrally distorted. Their exact structures are still not clear owing to the existing disorder [25]. (ii) The low-temperature form of [iso-$PrNH_3]_2[CuCl_4]$ contains nearly planar $[CuCl_4]^{2-}$ ions, which are weakly polymerized into a ribbon-like structure with 5- and 6-coordinate Cu(II) ions. The high-temperature form contains tetrahedrally distorted $[CuCl_4]^{2-}$ ions [26, 27]. The original explanation that a weakening of the network of hydrogen bonds brings about these structural changes and the accompanying thermochromism was, however, confirmed in each case [2].

It was also found that the complex $[NMe(C_2H_4Ph)H_2]_2$ $[CuCl_4]$, which shows the same type of green-yellow thermochromism, best fits the original view of Willet et al. In this case, discrete $[CuCl_4]^{2-}$ ions change their form from nearly perfect square planar to distorted tetrahedral (largest $\alpha = 138.1°$) upon heating [28]. Infrared, magnetic, NMR and ESR studies on these systems have yielded interesting data, but are often difficult to explain using a simplified model.

Other complexes, such as $[Me_2NH_2]_3[CuCl_4]Cl$ and $[MeNH_3]_2[CuCl_4]$, exhibit an apparently similar thermochromism with quite different spectral characteristics (cf. Fig. F.10). In these cases, the change in color does not take place abruptly, but over a large temperature range. The narrow bands observed at low temperatures are notably broadened and smeared together upon heating. Willett et al. concluded that this is due to an increase in molecular vibrations, which brings the energies of the electronic states gradually "out of focus". Such a spectral change is observed quite often when a colored solid is sufficiently heated (cf. Chap. F.I). In this case the ease of deformation of the $[CuCl_4]^{2-}$ ion may be the reason for this large spectral change [24].

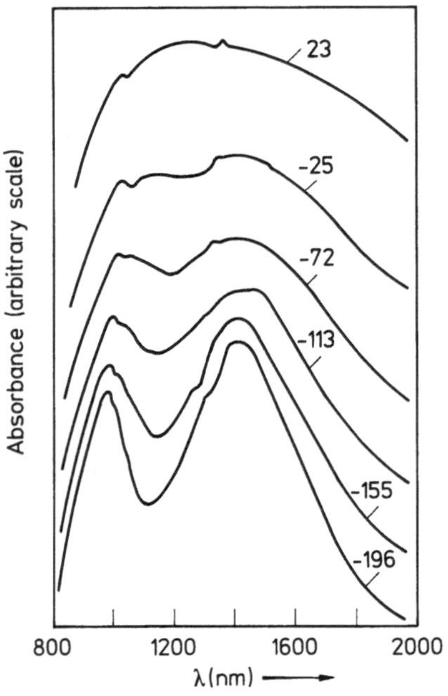

Fig. F-10. Thermochromic spectral changes of $[CuCl_4]^{2-}$ in the crystals of $[Me_2NH_2]_3$ $[CuCl_4]Cl$ sandwiched between glass plates. Successive curves are shown with vertically shifted scales. The numbers on curves are temp. (°C). (After Willett et al. [24])

Thermochromic transitions, which are essentially reversible, are also known among the so-called "tricholoro- and tribromo-cuprates", e.g. [iso-$PrNH_3]CuCl_3$ and [iso-$PrNH_3]CuBr_3$. The thermochromism of the former was discovered by Remy and Laves in 1933 [29] together with that of [iso-$PrNH_3]_2[CuCl_4]$. More detailed studies have only recently been carried out [30]. The $[CuCl_3]^-$ groups in [iso-$PrNH_3$]$CuCl_3$ are dimerized as $[Cu_2Cl_6]^{2-}$ ions:

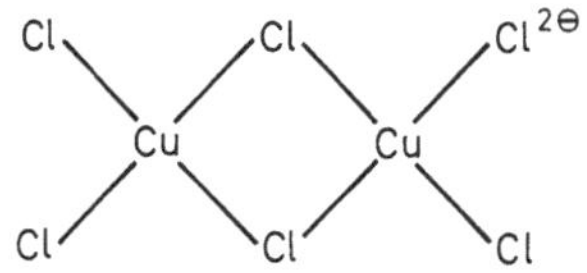

which are nearly planar. The dimers are weakly polymerized in a long chain-like structure. At 51 °C, the complex changes its color from brown to orange. The high-temperature modification contains a different chain of octahedral, 6-coordinate Cu(II)

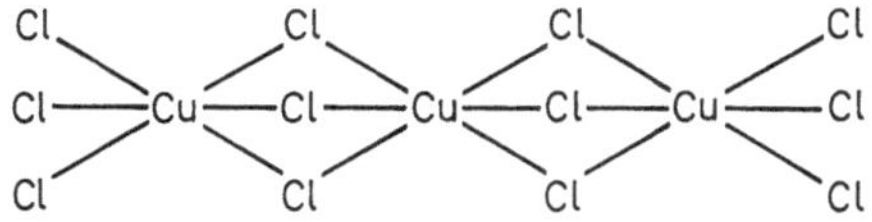

ions. A similar structural change was also observed with [iso-$PrNH_3$]$CuBr_3$ at 78 °C, although the complex has a very dark color (due to a CT band in the visible region) which makes an accurate observation of the accompanying change in color difficult. In this case, a second structural change takes place at 100 °C, which breaks down the triple bridges by thermal motion, forming singly bridged tetrahedral chains [31].

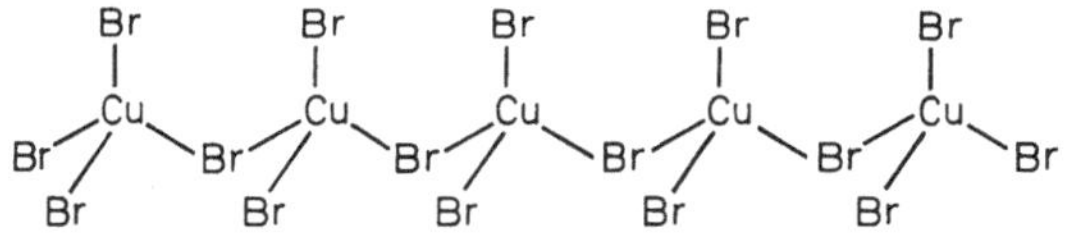

Further examples of reversibly thermochromic tetrahalogeno Cu(II) complexes, with somewhat different color changes, have been described by Marcotrigiano et al. [32]. They involve the salts of the piperazinium cation ($pipH_2^{2+}$) shown in Table F.4. The spectral studies again indicate that the structural changes are essentially of the (square planar → distorted tetrahedron) type, caused by a weakening of the network of hydrogen bonds.

Table F.4. Temperatures and color of thermochromic tetrahalo-Cu(II) complexes containing the piperazinium(2+) ion, $[H_2N(CH_2CH_2)_2NH_2]^{2+}$ = pip H_2^{2+}. (After Marcotrigiano et al. [32]).

Complex	Temp/°C	Color change
(pip $H_2)_2[CuCl_4]Cl_2$	80–95	Green-yellow → Bright yellow
(pip $H_2)_2[CuBr_4]Br_2$	80–95	Red-violet → Bright red
(pip $H_2)_2[CuBr_4]Cl_2$	80–95	Red → Bright red
(pip $H_2)_2[CuCl_3Br]Cl_2$	80–95	Red-orange → Bright red-orange

F.III.1.2 Thermochromism Associated with the Octahedral-Tetrahedral Changes of $[NiCl_4]^{2-}$ and Related Complexes

Drastic reversible (sometimes irreversible) thermochromism is often observed with the halide complexes of Ni(II). In most cases, the driving force seems to be similar to the one for the Cu(II) complexes. Goedken et al. [33] studied the 1,1,1-trimethylhydrazinium salt of $[NiCl_4]^{2-}$ from this point of view. At room temperature it is a yellow salt. It is transformed into a blue isomer at 145 °C. The latter slowly reverts back to the yellow form when it is cooled down below 120 °C. Prolonged heating above 220 °C brings about a disproportionation:

$$\underset{\text{blue}}{[Me_3N_2H_2]_2[NiCl_4]} \rightarrow \underset{\text{pinkish tan}}{[Me_3N_2H_2]NiCl_3} + [Me_3N_2H_2]Cl \qquad \text{(Eq. F.11)}$$

The $NiCl_3$-salt is of a polymeric structure (presumably octahedral). After cooling and standing the disproportionation mixture eventually reverts back to the yellow complex. The spectrum of the latter indicates that it has a distorted octahedral structure (remember that anhydrous $NiCl_2$ exhibits a $CdCl_2$ structure (C.N. = 6), and is also yellow), while the blue form was shown to contain tetrahedral $[NiCl_4]^{2-}$ ions. The situation is, therefore, quite similar to that in $[CuCl_4]^{2-}$. The main difference is the predominance of a 6-coordinate structure at room temperature in the former case instead of a square planar structure. The analogous bromide complex is also yellow and even more thermochromic, changing sharply to a blue color at 70 °C. In this case, however, the change was found to be irreversible, and cooling of the blue products leads to disproportionation.

A number of Ni(II) complexes of the general formula $(R_XNH_{4-X})_2[NiCl_4]$, where R is an alkyl or aryl group, were also found to be thermochromic, changing in the following way upon heating:

brown-yellow to greenish-yellow → blue to blue-green

The transition temperature varies considerably, from 70 °C to > 200 °C, depending upon R and x. It is higher for a complex containing a cation that is strongly hydrogen-bonding e.g., $MeNH_3^+$, than for a weak hydrogen-bonding cation such as Me_3NH^+. X-ray, spectral and magnetic measurements all support the view that the anion of the yellow form possesses a polymerized octahedral structure, while the blue form contains discrete $[NiCl_4]^{2-}$ anions [34]. When the cation is of the type $[RNH_3]^+$, it also seems that the thermochromism is really due to a disproportionation equilibrium in the solid state:

$$\underset{\text{yellow, octahedral}}{[RNH_3]NiCl_3 + RNH_3Cl} \underset{\text{Cool}}{\overset{\text{Heat}}{\rightleftharpoons}} \underset{\text{blue, tetrahedral}}{[RNH_3]_2[NiCl_4]} \qquad \text{(Eq. F.10)}$$

This is somewhat similar to the reaction observed with the hydrazinium complex [2].

Bloomquist and Willett summarized the thermochromic behavior of the Cu(II) and Ni(II) complexes of the type $[R_XNH_{4-X}]_2[MCl_4]$ in the following way [2]:

(1) In the Cu(II) complexes, the observed structural changes are square planar – distorted tetrahedral, while in the Ni(II) complexes they are distorted octahedral – tetrahedral.

(2) Changes in the Cu(II) complexes are favored when x > 1, while in the Ni(II) complexes they are favored when x > 2.

(3) In both cases, the strength of the network of hydrogen bonds determines the transition temperature.

F.III.2 Thermochromism of the Chelates of Copper(II) and Nickel(II) with Specific Stereochemistries

F.III.2.1 Thermochromism of the Copper(II) Chelates of the[CuN_4]-Type

Drastic reversible thermochromism in the solid state is often observed with the chelates of Cu(II) and Ni(II) possessing certain specific stereochemistries. A beautiful example of thermochromic Cu(II) chelate is given by [Cu(unsym-deen)$_2$](ClO_4)$_2$, discovered by Pfeiffer and Glaser in 1938 [35]. This chelate is red, unlike most of the alkyl derivatives of [Cu(en)$_2$](ClO_4)$_2$ which are violet of various shades, and changes reversibly at ca. 40 °C to bluish-violet.

This change in color was ascribed by Lever et al. [37, 38] to a structural change of the chelate cation from a square planar form to an elongated octahedral one, such as:

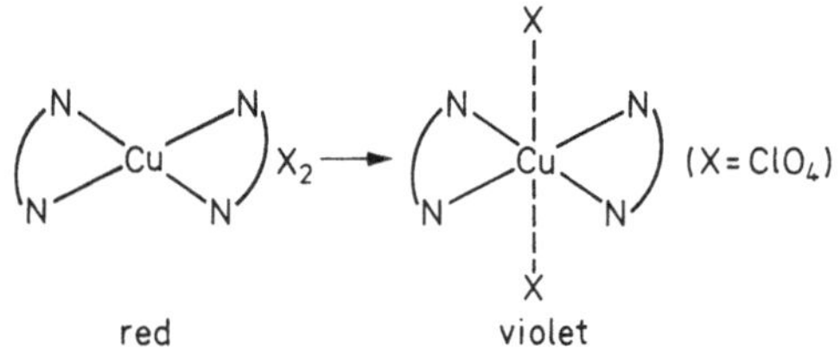

The color of a Cu(II) complex becomes more bluish, i.e., its absorption band is shifted to the red, as the interaction of the Cu(II) ion with the axial ligands increases. The thermochromic change was therefore explained in the following way: at room temperature, the axial positions of the Cu(II) ion are effectively shielded from outer influences by the bulky N-ethyl groups, so that the [CuN_4] chromophore of this complex can be treated as a perfectly square planar system without any axial ligands. So the d-d band is strongly shifted to the blue, as was shown in Chap. C, and the complex appears red. However, upon heating, the N-ethyl groups become increasingly mobile, so that the ClO_4^- ions can push them aside and get into the coordination sphere. This brings about a notable spectral shift to the red, making the complex appear blue. The bonding of the perchlorate ions to the Cu(II) ion is rather weak, so that they can easily be pushed back to their original positions by cooling the sample. A highly reversible color change thus comes about. Spectral data on this color change, which were first obtained by Hatfield et al. [36] and more recently by Yamaki and Fukuda [7a], seemed to be compatible with this view (cf. Fig. F.11).

Lever et al. studied the nature of this change in detail, comparing chelates with related ligands and different anions and using UV-visible, IR, magnetic and ESR technique [37, 38]. The results were apparently consistent with such a view. Fabbrizzi et al. studied these changes by means of spectral and thermal techniques, and obtained the results summarized in Fig. F.12 [39]. They pointed out that the chelate [Cu(unsym-deen)$_2$](BF_4)$_2$ exhibits a similar thermochromism at a lower temperature, while the corresponding Ni(II) chelates with ClO_4^- and BF_4^- change from yellow to red at around 100 °C (again the order of ease is $BF_4^- > ClO_4^-$). [Cu(unsym-deen)$_2$](NO_3)$_2$ also exhibits thermochromism at ca. 150 °C. Other salts of [Ni(unsym-deen)$_2$]$^{2+}$, i.e., those with NO_3^-, Br^-, 1/2 [$CdBr_4$]$^{2-}$ and I^-, only show a gradual color change towards orange, without a noticeable peak in DSC. They concluded that relatively small or easily deformable ions such as Br^-, I^-, etc. can easily approach the axial coordination sites of Ni^{2+} upon heating, so that a continuous thermochromism is

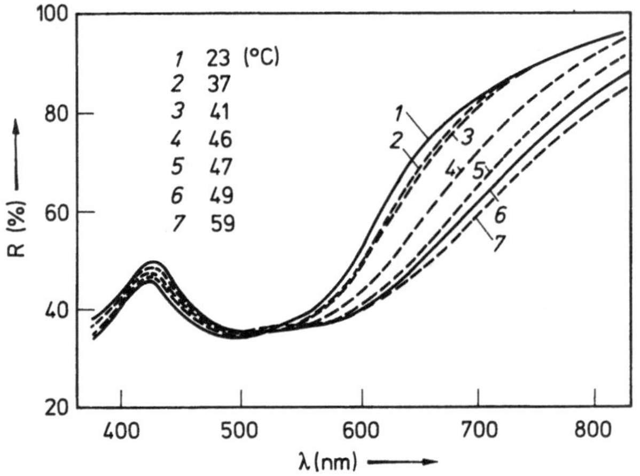

Fig. F-11. Thermochromic spectral change of [Cu(unsym-deen)$_2$](ClO$_4$)$_2$ measured with the device shown in Fig. F-4. The numbers on the curves are temp. (°C). (Taken from Yamaki and Fukuda [7a])

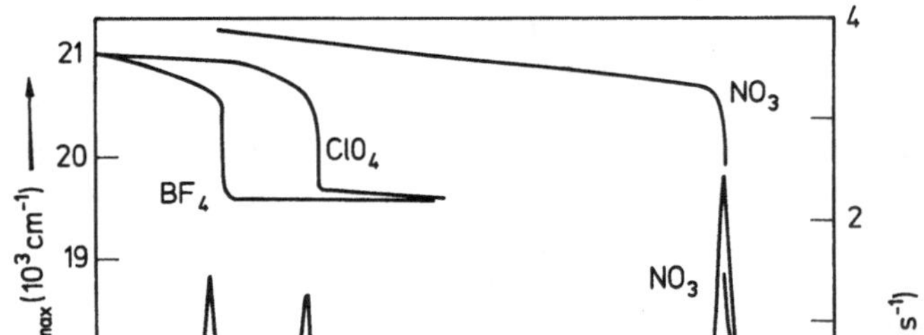

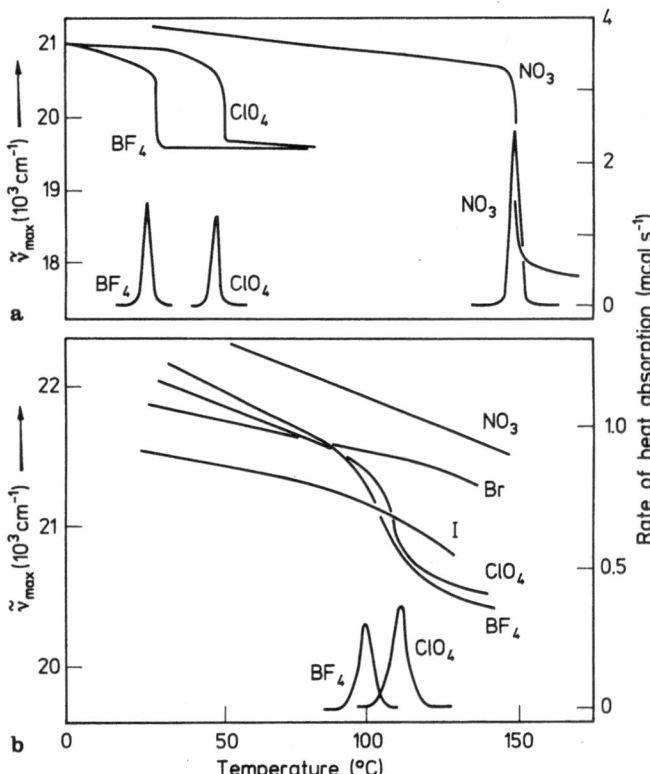

Fig. F-12 a, b. Changes in $\tilde{\nu}_{max}$ values of [Cu(unsym-deen)$_2$] (ClO$_4$)$_2$ **(a)** and [Ni(unsym-deen)$_2$] (ClO$_4$)$_2$ **(b)** with temperature. DSC curves of some complexes are also shown. (After Fabbrizzi et al. [39])

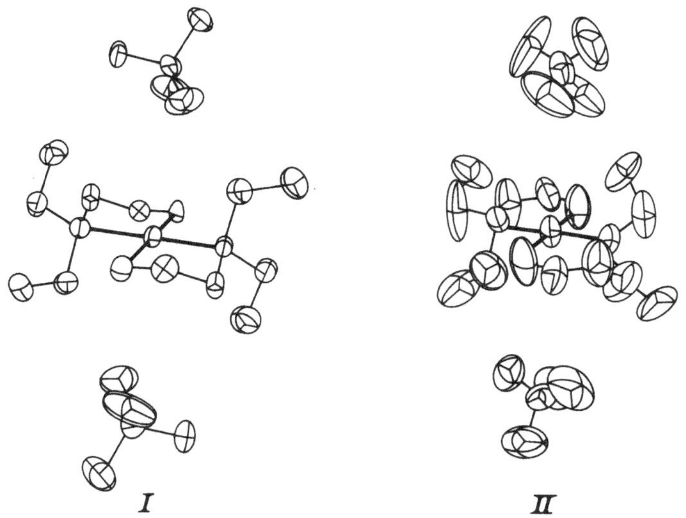

Fig. F-13. Structures of the red, low-temperature form (I) and the blue, high-temperature form (II) of [Cu(unsym-deen)$_2$] $(ClO_4)_2$. (After Grenthe et al. [40])

observed. Larger and less deformable ions such as BF_4^- or ClO_4^- need more activation energy in order to overcome the barrier produced by the bulky N-ethyl groups, so that a discontinuous thermochromism of the "transition type" results.

Ferraro et al. found that these thermochromic chelates are also piezochromic. In the case of the Cu(II) chelates, high pressures and high temperatures have the same effect. In the case of the Ni(II) chelates, the effect of high pressures is opposite, shifting the band to the blue [39a–b]. These data were also considered to be compatible with Lever's view.

However, many of these earlier ideas have to be reconsidered, in view of the X-ray studies of Grenthe et al. [40] on the complex, [Cu(unsysm-deen)$_2$]$(ClO_4)_2$. Figure F.13 illustrates their results. It can be seen that the tetragonalities of Cu^{2+} in both the red and violet forms (I and II) are not very different. The main difference lies in the conformation of the en rings. The conformation in the red form is gauche, which is common in most chelates containing en; in the violet form, on the other hand, the chelate rings are apparently planar with diminished C-C bonding distances (from 151 pm in I to 137 pm inII) and large vibrational amplitudes perpendicular to the ring.

These features indicate that the increase in the vibration weakens the network of hydrogen bonds in the crystal, and brings about a rapid conformational change in the chelate rings above a certain temperature. Only the average of such constantly changing structures is reflected in Fig. F.13. An ideal overlap of metal and ligand orbitals is thus disturbed by such "rattling", so that the apparent ligand field strength of the diamine is weakened and the d-d band is shifted towards the red. It must be mentioned that Hatfield et al. [36], who are pioneers in this field, had suggested such a mechanism a long time ago, without receiving much attention.

Another dramatically thermochromic Cu(II) chelate, [Cu(daco)$_2$]$(NO_3)_2$, was prepared by Yamaki et al. [7]. It has the structure shown in Fig. F.14. The chelate is orange at room temperature, but turns reversibly violet at ca. 90° C (cf. Fig. F.15). In

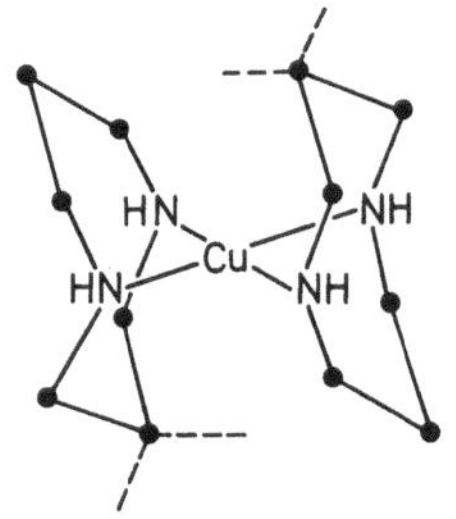

Fig. F-14. Structure of $[Cu(daco)_2]^{2+}$ in $[Cu(daco)_2](NO_3)_2$. • = CH_2 (After Yamaki et al. [7])

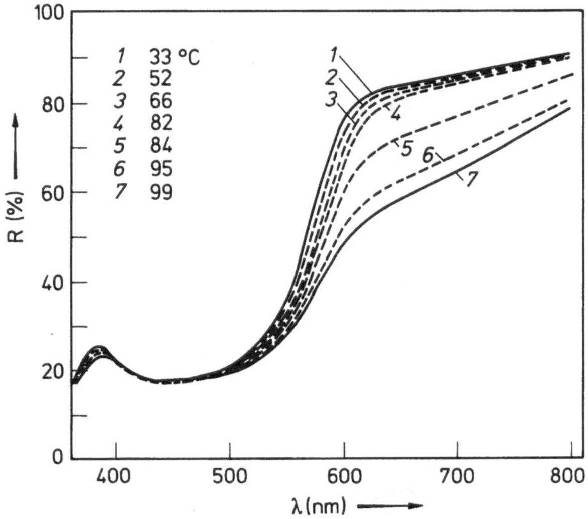

Fig. F-15. Thermochromic spectral changes of $[Cu(daco)_2](NO_3)_2$ measured with the device shown in Fig. F-4. (Taken from Yamaki et al. [7])

this case, however, the reverse change in color caused by cooling is often slow, so that a large hysteresis is observed. The X-ray structure of the orange form is now known. It shows that the axial positions of the Cu(II) ion are practically vacant. The NO_3^- ions lie adjacent to the chelate cation, forming hydrogen bonds with its two NH groups. The structure of the violet form is still unknown, but IR data indicate that the NO_3^- ions in it are weakly coordinated to the Cu(II) ion as monodentate ligands.

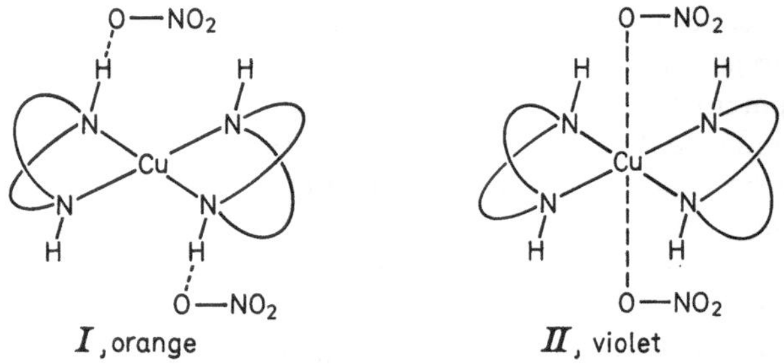

In the orange form (I), therefore, the Cu^{2+} ion lies in a completely square planar ligand field; here two of the Cu-N bonds are strengthened by the hydrogen bonds, which polarize the NH groups and increase the charge on the nitrogen atoms. In the violet form (II), on the other hand, the Cu^{2+} ion lies in an elongated octahedral field with two NO_3^- ions at the axial positions, and its equatorial field is somewhat weakened by the breakdown of the hydrogen bonds. These changes, which are brought about by the thermal motions of the ligands, are probably the cause of the drastic thermochromism.

It may be mentioned that a different type of thermochromism for Co(II) complexes of daco was reported by Musker and Steffen [41]. For example, $Co(daco)_2Cl_2$ and $Co(daco)_2Br_2$ are both pale pink complexes that lose a daco ligand at 60–64 °C to form blue $Co(daco)X_2$. On the other hand, $Co(daco)_2I_2$.EtOH is pale orange and changes reversibly into a pink form at 75 °C. These bis chelates are seemingly square pyramidal 5-coordinate ones, $[Co(daco)_2X]X$. The structure of the pink modification of the iodide complex is still not known.

F.III.2.2 Thermochromism of the Nickel(II) Complexes of Some Heterocyclic N-ligands Due to Monomer-Dimer Equilibria

Another kind of reversible thermochromism of metallic chelates, which has been studied by a large number of investigators, is due to a monomer-dimer equilibrium for certain Ni(II) chelates:

$[NiLCl_2]_2$	$\rightleftharpoons$	$2[NiLCl_2]$	(Eq. F.13)
dimer, yellow or orange, square pyramidal		monomer, violet or purple, distorted tetrahedral	

The change takes place at a certain transition temperature. The reverse change is usually slow, showing a large hysteresis as in the case of $[Cu(daco)_2](NO_3)_2$. The ligands L are usually bulky bidentate diamines with one or both of the nitrogen atoms involved in heterocyclic π-system(s):

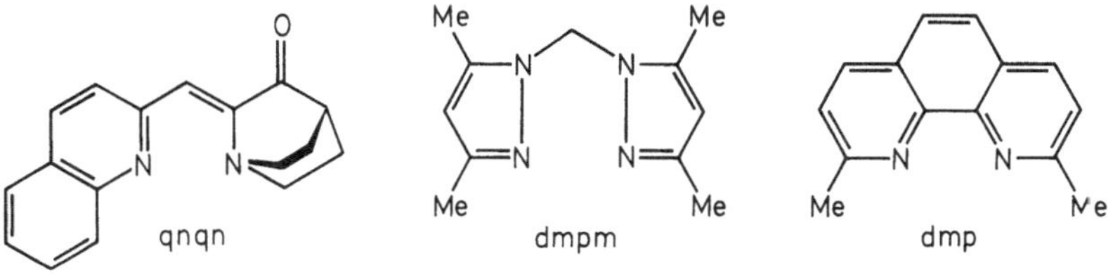

The names of these ligands are: qnqn = trans-2-(2′-quinolyl)-methylene-3-quinuclidinone, dmpm = bis(3,5-dimethylpyrazolyl)-methane, and dmp = 2,9-dimethyl-1,10-phenanthroline.

The complex $Ni(qnqn)Cl_2$ [42–44] consists of two isomers; one is violet, the other yellow. The former is obtained by heating $NiCl_2.6H_2O$ and qnqn in an alcoholic medium and cooling rapidly with ice. The latter is prepared by letting a solution of the violet form in CH_2Cl_2 evaporate at room temperature. Both are stable at room temperature and can be stored for a long time. However, when the yellow isomer is heated gradually, it changes into the violet isomer at 230 ± 10 °C. The yellow color returns upon cooling, but the temperature must be lowered down to at least −78 ± 8 °C for a noticeable change to occur. The value of ΔH for the dimerization reaction was estimated to be $-10.9\,kJ\,mol^{-1}$.

The structure of the yellow dimer is now known. It is a chloride-doubly-bridged dimer of square pyramidal structure. The structure of the violet monomer is assumed to be that of a distorted tetrahedron, similar to that of the Co(II) chelate. Thus it is clear that the double bridge is broken down by heating up to 230 °C. The resulting monomer contains qnqn in a highly planar form, while in the dimer it is markedly non-planar.

Thus, the reverse transition, monomer-dimer, is strongly hindered by the difficulties involved in achieving the large structural change in qnqn and a favorable reorientation of the monomer units. Once formed, the violet form remains in a metastable state down to ca. $-80\,°C$, at which point the decreases in the size of the unit cell and in the vibrations begin to force the monomers into the stable dimeric form.

These observations lead to the idea that the dimerization is facilitated at high pressures, i.e., the system should also be piezochromic. In fact, Long and Ferraro [45] have shown that the violet isomer is irreversibly converted to the yellow isomer under a pressure of ca. 2 kbar. A similar piezochromic change was also observed with the Co(II) complex, which served as the model for the violet isomer.

The orange crystals obtained in the reaction of $NiCl_2.6H_2O$ with dmpm in acetone adopt a deep purple color above 220 °C (ΔH is ca. $-10.5\,kJ\,mol^{-1}$). Cooling down to 130 °C restores the orange color. Spectral, magnetic and IR studies indicate that the structural change is very similar to the one for the qnqn chelate [46].

The purple and yellow forms of the dmp chelate can be prepared as follows:

$$\left.\begin{matrix} NiCl_2.6H_2O \\ dmp \end{matrix}\right\} \xrightarrow{\text{mix in EtOH}} \underset{\text{green}}{Ni(dmp)Cl_2.H_2O} \xrightarrow{\text{heat}} \underset{\text{purple}}{Ni(dmp)Cl_2} \qquad \text{(Eq. F.14)}$$

$$Ni(dmp)Cl_2.H_2O \xrightarrow{\text{evaporation of } CHCl_3 \text{ solution}} \underset{\text{yellow}}{Ni(dmp)Cl_2}$$

The yellow isomer is abruptly changed to purple at 180° C. In this case, the reverse change is especially hard to bring about. Even cooling in liquid N_2 is not effective. The essential similarity between this change and those of the qnqn and dmpm chelates is evident from various structural studies. The difficulty of the reverse reaction is ascribed to the rigidity of the bulky dmp ligand, which hinders a rearrangement of the monomer to the dimer [44, 47, 48].

As a whole, the thermochromism of these chelates may be viewed as a compromise undergone by the central Ni(II) ion. This ion exhibits a strong tendency for octahedral coordination and, when the ligand field strength is high enough, square planar (low-spin) coordination. In the present case, however, the bulkiness of the chelating ligand and the relative weakness of the Cl^- ligands force the complex to adopt a less favorable structure. A tetrahedral one is the most free from steric hindrance, but the accompanying ligand field will be too weak. The formation of a square pyramidal dimer will lead to a higher ligand field strength, but the steric repulsion also increases. Thus, the thermochromic change is essentially a struggle between enthalpy (ligand field strength) and entropy(steric) factors.

F.III.3 Miscellaneous Examples of Solid-State Thermochromism

Dinitrodiamminecopper(II), $[Cu(NO_2)_2(NH_3)_2]$, was first prepared in 1922 by heating $[Cu(NH_3)_4](NO_2)_2$ [49]. It is purple at temperatures below 31–32 °C, but green above this point. Mori et al. [50] found that the transition temperature can be lowered down to ca. $-50\,°C$ by replacing a fraction of the $NO_2{}^-$ ions with Cl^- or Br^- ions. Such is the case for $CuCl_{0.3}(NO_2)_{1.7}(NH_3)_2$ and $CuBr_{0.23}(NO_2)_{1.77}(NH_3)_2$. When the concentration of Cl^- or Br^- ions is still higher (> 6% by weight for Cl^- and > 10% for Br^-), the green form becomes stable at all temperatures above that of liquid nitrogen.

Magnetic, DSC and X-ray studies of these complexes have proven that the thermochromic changes involve transitions between two isomeric forms of the same composition:

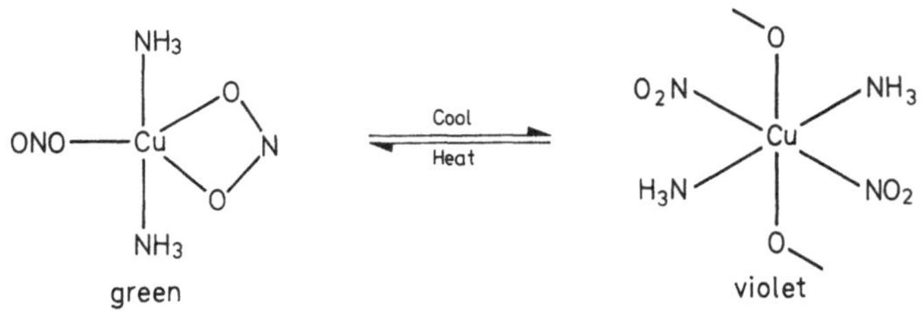

The structure of the violet form is easy to understand. It consits of a common elongated tetragonal complex of Cu(II), in which the NO_2^- ions are equatorially coordinated to the Cu(II) ion via their nitrogen atoms and form additional axial bonds with adjacent complexes. The green form, on the other hand, is made up of trigonal-bipyramidal units with O-coordinated NO_2^- ions, one of which is chelated while the other is not. This latter NO_2^- ion is the one that may be substituted by a halide ion, which hinders the transition to the violet form.

Another kind of complex, for which the ease of the thermochromic change can be "tuned" by mere changing of composition, was prepared by Bereman and Brubaker [51]. They studied a series of V(III) complexes ($3d^2$) with the composition $[Et_4N][VX_4(CH_3CN)_2]$, where X^- is Cl^- or Br^-. Both halide ions may be in the same crystal, i.e., X_4 may be Cl_3Br, Cl_2Br_2 or $ClBr_3$. When $X_4 = Cl_4$, the complex is yellow at room temperature, 195K and 77K. When $X_4 = Br_4$, the complex is reddish brown at room temperature, but yellow at lower temperatures. Complexes containing both halide ions become more and more thermochromic as the amount of Br^- increases. The spectral data indicate that all of the complexes that have been studied are in fact thermochromic, due to the shift of their strong CT bands in the UV or blue part of the spectrum towards shorter wavelengths and their remarkable sharpening brought about by a decrease in temperature. When the complex is rich in Cl^- ions, these spectral changes take place with color changes that are scarcely perceptible to the naked eye, so that an apparently large difference between Cl^--rich and Br^--rich complexes comes about.

Still more examples of reversible thermochromism with metallic complexes can be found in the Reference [1, 2, 52]. Even simple or double oxides or halides of certain heavy metals (especially iodides of Hg^{2+} and Ag^+) can sometimes be strongly thermochromic (cf. Chap. A.II.3). K_2CrO_4 shows a drastic change, from yellow to orange and then to red, upon heating up to its melting point (975 °C). It reverts to yellow upon cooling [53a]. Even ZnO is known to become reversibly yellow upon heating to ca. 300 °C. CdO undergoes various changes in color from greenish-yellow to brown, and sometimes nearly black, depending on its thermal history. The colors of these oxides are ascribed to the formation of crystal defects and are probably also connected with the fact that these oxides are, by no means, simple ionic crystals, but are markedly covalent with semiconducting characteristics [53]. S_4N_4 is not a metallic compound, but is also extremely thermochromic. Its orange color fades upon cooling (pale yellow below ca. −30 °C) and turns deep red at 100 °C [53b].

It may be added that certain glasses have been reported to be thermochromic. For example, glasses with reduced phosphate in $K_2O—B_2O_3—Al_2O_3—P_2O_5$ systems can either be colorless or red according to their previous heat treatment, while $Na_2O—B_2O_3—MgO—Al_2O_3$ glasses containing Bi_2O_3 (originally golden yellow) become reversibly dark upon heating. Although the underlying principles of these changes are still not clear, such glasses may be of practical interest in the future [54, 55].

F.III.4 Fluorescence Thermochromism of Cu(I) Complexes

Fluorescence thermochromism is an interesting phenomenon, that has been extensively studied by Hardt et al. [56, 57]. Cu(I) compounds, being d^{10}, are usually colorless and not thermochromic in the usual sense. However, when CuI.py is irradiated by UV radiation, it emits a beautiful yellow fluorescence at room temperature which becomes violet at 77K. The appearance of this fluorescence depends intimately on the structure of the compound. The fluorescence is green with $CuI.py_2$ irrespective of the temperature, while $CuI.py_3$ does not emit any fluorescence at all. Some of the results [56, 57] on various Cu(I) compounds of the formula CuLX ($X^- = Cl^-$, Br^- or I^-, L = organic base) are summarized in Table F.5.

Table F.5. Fluorescence thermochromism of CuLX. [After Hardt (56, 57)]

		Color change	
L	X	Room temp. (°C)	Low temp. (°C)
NH_3	I	Orange	Red (−180)
CH_3NH_2	I	Rose	Red (−180)
$(CH_3)_2NH$	I	Orange	Red (−180)
$(CH_3)_3N$	I	Yellow	Red (−180)
$(C_2H_5)NH_2$	I	Orange	Blue-violet (−180)
py	I	Yellow	Red (−180); Violet (−196)
2-Mepy	Br	Red	Blue (−196)
2-Mepy	I	Bright red	Dark blue (−196)
3-Mepy	I	Yellow	Sky blue (−196)
4-Mepy	Cl	Red	Turquoise (−196)
2,4-Me_2py	Br	Orange	Blue (−180)
2,4-Me_2py	I	Ocher	Blue (−180)
4-NH_2py	Br	Rose	Blue (−196)
Quinoline	I	Orange	Orange (−180)

Many other examples are known. Although the mechanism of this phenomenon is still unknown, Hardt et al. have proposed analytical applications. For example, various derivatives of pyridine can be distinguished from each other by spotting them on a piece of paper impregnated with CuI and irradiating this paper with a UV lamp, first at room temperature and then at the temperature of liquid N_2 [58, 59]. So far no other metal complexes are known to show this kind of thermochromism.

F.IV Practical Applications of Solid-State Thermochromism: Color Indicators for Temperature Changes

Before ending this chapter, mention will be made of the practical application of these phenomena. Metal compounds which exhibit reversible or irreversible thermochromism are being widely used in the manufacture of color indicators for temperature changes. The latter are sold in the form of paints, printing and writing inks (often with specially designed pens), crayons, or paper or plastic adhesive lables. The temperature at which a change in color is observed can vary widely between 0 °C and above 1000 °C, depending on the choice of materials and methods of preparation. By painting, marking, or pasting them, and watching their colors, one can readily see the temperatures of various part of an engine, a furnace, a reaction tower, or the knife-edge of a cutting tool, etc., under operating conditions, and keep a record by taking colored pictures. Domestic products, such as bath or kitchen thermometers, or pitchers or glasses which change color when hot or cold drinks are inside, are also on the market.

Table F.6. Examples of thermochromic compounds used as temperature indicators.

A: Irreversible systems[a, b] (After Cowling et al. [60])

Compound	Color change		Temp./°C
	Low temp.	High temp.	
CoI_2. 2hmta. $8H_2O$[c]	Brown pink	Green	50
$NiBr_2$. 2hmta. $9H_2O$	Green	Blue	62
$Co(AcO)_2$. $4H_2O$[c]	Pink	Purple	82
$Co(NCS)_2$. 2py. $10H_2O$	Lavender	Blue	93
$CoSiF_6$	Orange pink	Bright pink	99
$Co(HCO_3)_2$. $2H_2O$[c]	Pink	Deep lavender	116
[Cr en_3]Cl_3 (I)[d]	Yellow	Red	119
[Cr en_3]$(NCS)_3$ (I)	Yellow	Red	121
NH_4VO_3 (I)	White	Pink	132
NH_4VO_3 (II)[e]	Pink	Black	164
[Cr en_3]$(NCS)_3$ (II)	Red	Black	252
[Cr en_3]Cl_3 (II)	Red	Black	270

[a] In some cases, the original color reverts after long standing.
[b] Each change was observed on a paint which contains the compound, 30% methacrylate solution, dibutyl phthalate, lacquer thinner and TiO_2 in the ratio of 25:50:25:20:10.
[c] Number of crystal water given by Cowling et al. was corrected.
[d] First change.
[e] Second change.

B: Reversible systems[a] (After Halmos and Wendlandt [61])

Compound	Color change		Temp.[b]/°C
	Low temp.	High temp.	
Ag_2HgI_4	Yellow	Orange	47– 50
Cu_2HgI_4	Red	Brown black	70– 71
$PbHgI_4$	Orange red	Yellow	129 – 135
HgI_2	Red	Yellow	130–133
AgI	Yellow	Brown	144–145

[a] Owing to hysteresis, the backward change may be slow, and the temperature for it may be considerably lower.
[b] Determined by EC(electric conductivity)-DTA methods.

Though numerous patents hinder a close study of the materials used in these commercial products, many of them seem to be complexes of the first transition series (especially Co(II) or Co(III) complexes) which often exhibit irreversible thermochromism, or simple or double iodides of Ag^+ or Hg^{2+} which show reversible thermochromism. Many such examples are found in the reviews by Day [1, 52] and, in particular, in the reports by Gvozdov and Erunova [4], Cowling et al. [60], Halmos and Wendlandt [61]. Some are listed in Table F.6A and B.

Mixing these materials with other thermochromic or nonthermochromic materials may considerably change the mode and the ease of the color change. In some cases stepwise color changes may be induced (even a single compound can show such an effect; cf. Table F.6A). The temperatures shown by these indicators are only approximate, so that proper care must be taken if an accuracy within ca. $\pm 5\%$ is desired.

In recent years, thermochromic products that make use of organic liquid crystals, which often exhibit marvellous reversible color changes for slight changes in temperature, have appeared on the market. This very interesting topic is, however, outside the scope of this book.

References

1. Day, J. H.: Chem. Rev. **68**, 649 (1968)
2. Bloomquist, D. R., Willett, R. D.: Coord. Chem. Rev. **47**,125 (1982)
3. Wilke, K. T., Opfermann, W.: Z. phys. Chem. (Leipzig) **224**, 237 (1963)
4. Gvozdov, S. P., Erunova, A. A.: Khim. Khim. Tekhnol. **5**, 154 (1958)
5. Wendlandt, W. W.: "Thermal Methods of Analysis", Interscience-Wiley, New York (1964)
6. Wendlandt, W. W., Smith, J. P.: "Thermal Properties of Transition Metal Ammine Complexes", Elsevier, Amsterdam (1967)
7. Yamaki, S. et al.: Chem. Lett. **1982**, 269

7a. Fukuda, Y., Yamaki, S.: Hitachi Sci. Instrument News **24**, 9 (1981)

8. Nagase, K. et al.: Thermochim. Acta **23**, 283 (1978)
9. Nagase, K. et al.: ibid. **31**, 391 (1979)
10. Barbieri, G. A., Calzolari, F.: Atti Accad. Lincei **19 II**, 584 (1910)
11. Nagase, K. et al.: Bull. Chem. Soc. Jpn. **49**, 1563 (1976)
12. Ihara, Y. et al.: Bull. Chem. Soc. Jpn. **55**, 1028 (1982)
13. Tsuchiya, R. et al.: Bull. Chem. Soc. Jpn. **55**, 1858 (1982)

13a. Ihara, Y. et al.: Thermochim. Acta **67**, 23 (1983)

13b. Ihara, Y., Tsuchiya, R.: Bull. Chem. Soc. Jpn. **57**, 2829 (1984)

13c. Ihara, Y.: Bull. Chem. Soc. Jpn. **58**, 3248 (1985)

13d. Ihara, Y. et al: Bull. Chem. Soc. Jpn. **59**, 2309 (1986)

13e. Related data can be found in: Saito, R., Kidani, Y.: Bull. Chem. Soc. Jpn. **52**, 2320 (1979)

14. Tsuchiya, R. et al.: Chem. Lett. **1976**, 911
15. Ihara, Y., Tsuchiya, R.: Bull. Chem. Soc. Jpn. **53**, 1614 (1980)

15a. Ferraro, J. R. et al: J. Am. Chem. Soc. **93**, 3862 (1971)

15b. Zahner, J. C., Drickamer, H. G.: J. Chem. Phys. **33**, 1625 (1960)

16. Uehara, A. et al.: Inorg. Chem. **21**, 2422 (1982)
17. Akabori, K., Kushi, Y.: J. Inorg. Nucl. Chem. **40**, 625 (1978)
18. Akabori, K., Kushi, Y.: ibid. **40**, 1317 (1978)
19. Tsuchiya, R. Uehara, A.: Thermochim. Acta **50**, 93 (1981)
20. Tsuchiya, R. et al.: Inorg. Chem. **21**, 590 (1982)
21. Ueno, H.: Bull. Chem. Soc. Jpn. **54**, 1821 (1981)

22. Tsuchiya, R. et al.: Bull. Chem. Soc. Jpn. **53**, 921 (1981)
23. Tsuchiya, R. et al.: ibid. **55**, 3770 (1982)
24. Willett, R. D. et al.: Inorg. Chem. **13**, 2510 (1974)
25. Simonsen, S. H.: cf. ref. [2]
26. Anderson, D. N., Willett, R. D.: Inorg. Chim. Acta **8**, 167 (1974)
27. Bloomquist, D. R., Willett, R. D.: J. Am. Chem. Soc. **103**, 2615 (1981)
28. Harlow, R. L. et al.: Inorg. Chem. **13**, 2106 (1974)
29. Remy, H., Laves, G.: Ber. Bunsengesell. phys. Chem. **66**, 401 (1933)
30. Roberts, S. A. et al.: J. Am. Chem. Soc. **103**, 2603 (1981)
31. Bloomquist, D. R.: J. Am. Chem. Soc. **103**, 2610 (1981)
32. Marcotrigiano, G. et al.: Inorg. Chem. **15**, 2333 (1976)
33. Goedken, V. L. et al.: Inorg. Chem. **10**, 2682 (1971)
34. Ferraro, J. R., Sherren, A. T.: Inorg. Chem. **17**, 2498 (1978)
35. Pfeiffer, P., Glasser, H.: J. prakt. Chem. **151**, 134 (1938)
36. Hatfield, W. E. et al.: Inorg. Chem. **2**, 629 (1964)
37. Lever, A. B. P., Mantovani, E.: Inorg. Chem. **10**, 817 (1971)
38. Lever, A. B. P. et al.: Inorg. Chem. **10**, 2424 (1971)
39. Fabbrizzi, L. et al.: Inorg. Chem. **13**, 3019 (1974)
39a. Ferraro, J. R. et al.: Inorg. Chem. **15**, 2342 (1976)
39b. Ferraro, J. R. et al.: Inorg. Chem. **16**, 2127 (1977)
40. Grenthe, I. et al.: Inorg. Chem. **18**, 2687 (1979)
41. Musker, W. K., Steffen, E. D.: Inorg. Chem. **13**, 1951 (1974)
42. Long, G. J., Coffen, D. L.: Inorg. Chem. **13**, 270 (1974)
43. Long, G. J. et al.: J. Chem. Soc., Dalton Trans. **1975**, 762
44. Laskowski, E. J. et al.: Inorg. Chem. **15**, 2908 (1976)
45. Long, G. J., Ferraro, J. R.: J. Chem. Soc. Chem. Commun. **1973**, 719
46. Reedijk, J., Verbiest, J.: Trans. Metal. Chem. **3**, 51 (1978)
47. Preston, H. S. et al.: J. Inorg. Nucl. Chem. **30**, 1463 (1968)
48. Preston, H. S., Kennard, C. H. L.: J. Chem. Soc. **A1969**, 2682
49. Bassett, H., Durrant, R. G.: J. Chem. Soc. **1922**, 2630
50. Mori, M. et al.: Inorg. Chem. **14**, 1002 (1975)
51. Bereman, R. D., Brubaker, C. H., Jr.: J. Inorg. Nucl. Chem. **32**, 2557 (1970)
52. Day, J. H.: "Chromogenic Materials", in Kirk-Othmer: "Encyclopedia of Chemical Technology", Vol. 6, Wiley-Interscience, New York (1979), p. 129
53. Coogan, C. K., Rees, A. L. G.: J. Chem. Phys. **20**, 1650 (1952)
53a. Uemura, T. et al.: Nippon Kagaku Zasshi **74**, 670 (1953)
53b. Greenwood, N. N., Earnshaw, A: "Chemistry of the Elements", Pergamon, Oxford (1984) p. 856
54. Abe, Y. et al.: J. Am. Ceram. Soc. **64**, 206 (1981)
55. Sen, A. et al.: Materials Sci. Lett. **2**, 677 (1983)
56. Hardt, H. D.: Naturwissenschaften **61**, 107 (1974)
57. Hardt, H. D., Gechnizdjani, H.: Inorg. Chim. Acta **15**, 47 (1975)
58. Hardt, H. D., Pierre, A.: Fresenius Z. Anal. Chem. **265**, 337 (1973)
59. Wagner, H., Lehmann, H.: Fresenius Z. Anal. Chem. **283**, 115 (1977)
60. Cowling, J. E. et al.: Ind. Eng. Chem. **45**, 2317 (1953)
61. Halmos, Z., Wendlandt, W. W.: Thermochim. Acta **7**, 113 (1973)

Concluding Remarks

The readers who have had the patience of reading through this book have probably recognized an important fact, i.e., the colors or spectra of inorganic substances are by no means invariable characteristics, but are susceptible to various outer influences, sometimes gradually at other times abruptly, sometimes physically at other times chemically. Careful studies of the phenomena, i.e., thermochromism, piezochromism and solvatochromism (and also other "chromisms", e.g., photochromism and electrochromism, which are not covered in this book), can therefore lead to a better understanding of the electronic structures of these substances and their changes during phase transitions, chemical equilibria or reactions, in nearly every aspect of present-day inorganic chemistry and related fields.

In spite of the intrinsic importance and interest in these phenomena, we do not know much about them. For the most part, the data is a collection of fragmentary observations obtained by chance in the course of different studies. As the number of known inorganic compounds (especially metallic complexes and chelates) is really vast, and is still increasing every day, it is highly probable that a large number will exhibit interesting chromotropic behavior on closer examination. It will also be possible, with some good ideas and some skill, to design and synthesize compounds with certain chromotropic properties. Such compounds may find many practical uses, in technology, in education, and as models in bio-scientific research.

These expectations seem to promise a fruitful future for ambitious and adventurous chemists searching for new subjects of study in inorganic chemistry. The sussessful completion of such studies will, however, be hard work. A number of new techniques and theories must be developed simultaneously in order to keep pace with the novel compounds and phenomena. We shall repeatedly come across the never-ending paradox "if one does not know everything, one can not know anything at all." However, the fact that chemistry is such a beautiful science will remain as our permanent relief.

Subject Index

Thermochromism, piezochromism and solvatochromism are denoted by therm., piez. and solvat., respectively.